# HUMAN INTERACTION WITH MACHINES

# Human Interaction with Machines

## Proceedings of the 6th International Workshop held at the Shanghai JiaoTong University, March 15-16, 2005

*Edited by*

G. HOMMEL
*Technische Universität Berlin,
Germany*

and

SHENG HUANYE
*Shanghai Jiao Tong University,
Shanghai, PR China*

Springer

A C.I.P. Catalogue record for this book is available from the Library of Congress.

ISBN-13 978-90-481-7021-0
ISBN-10 1-4020-4043-1 ( e-book)
ISBN-13 978-1-4020-4043-6 (e-book)

---

Published by Springer,
P.O. Box 17, 3300 AA Dordrecht, The Netherlands.

*www.springer.com*

*Printed on acid-free paper*



Printed in the Netherlands.

# Contents

# Committee

## Workshop Co-Chairs

Günter Hommel
Sheng Huanye

## Program Committee

Fu Yuxi
Günter Hommel
Sheng Huanye

## Organizing Committee

Fu Yuxi (Chair)
Gu Lixu
Li Fang
Lu Baoliang
Ma Lizhunag
Yao Tianfang
Zhang Liqing

## Editorial Office and Layout

Wolfgang Brandenburg

# Preface

The International Workshop on "Human Interaction with Machines" is the sixth in a successful series of workshops that were established by Shanghai Jiao Tong University and Technische Universität Berlin. The goal of those workshops is to bring together researchers from both universities in order to present research results to an international community.

The series of workshops started in 1990 with the International Workshop on "Artificial Intelligence" and was continued with the International Workshop on "Advanced Software Technology" in 1994. Both workshops have been hosted by Shanghai Jiaotong University. In 1998 the third workshop took place in Berlin. This International Workshop on "Communication Based Systems" was essentially based on results from the Graduiertenkolleg on Communication Based Systems that was funded by the German Research Society (DFG) from 1991 to 2000. The fourth International Workshop on "Robotics and its Applications" was held in Shanghai in 2000. The fifth International Workshop on "The Internet Challenge: Technology and Applications" was hosted by TU Berlin in 2002.

The subject of this year's workshop has been chosen because both universities have recognized that human interaction with machines in different application fields has become a major issue. Highly sophisticated devices like video recorders, mobile phones, digital cameras, and all the nice infotainment equipment in modern cars offer a great amount of functionality and comfort – but almost nobody is really able to use more than a tiny part of their functionality. The manuals to operate those devices are often bigger in size and weight than the devices themselves. The same applies to services that are offered in the web or by telecommunication providers. Only few persons even know about the complex features offered by those systems. Similar and even more difficult problems arise if we think of robots or medical devices that should be used as helpmates directly in contact with handicapped or elderly persons.

To overcome those problems intelligent systems have to be designed that derive their behavior at least partially by observing human behavior, learning from it, but also guaranteeing safety requirements. So the interfaces must take into account more modalities than currently involved in human machine interaction as e.g. visual information, natural language, gestures, but also physiological data like EEG, ECG, and EMG.

All those aspects of human interaction or even cooperation with machines are addressed in this workshop. The continuous support of both universities is gratefully recognized that enabled the permanent exchange of ideas between researchers of our two universities.

Shanghai, March 2005

Günter Hommel and Sheng Huanye

# Reference Points for the Observation of Systems Behaviour

Sebastian Bab and Bernd Mahr

bab@cs.tu-berlin.de mahr@cs.tu-berlin.de
Technische Universität Berlin, Germany

**Abstract.** Motivated from the notion of reference points in the ISO standard on Open Distributed Processing (RM-ODP) the question is examined what meaning a general concept of reference points in the context of complex systems and their development could be given. The paper discusses three approaches for the observation of systems behaviour and draws some conclusions in view of such concept.

## 1 Introduction

Human interaction with machines is not only the use of the systems functionality as it is designed to fulfill certain user oriented purposes and intentions, but also concerns interaction for the purpose of management, control, verification, testing, and monitoring of its behaviour. The goal of this paper is to discuss the concept of reference points as a means for the observation of systems behaviour.

The paper is pointing at a rather open field of questions. These questions result from the fact that systems which attract today's research interest, have characteristics that make specification and verification a highly complicated task. Among those characteristics are largeness (like the world wide web), evolution in time (like information and communication systems in large institutions), complexity due to the necessary decomposition into large numbers of subsystems or components (like today's operating systems), integration due to the need for interoperability with pre-existing subsystems or components (like services in the context of e-government), embeddedness (like software in the automotive domain), or autonomy due to multiple domains in the systems application structure (like patient records in medicine, in a network of specialized clinics, hospitals and practitioners). For the task of developing such systems numerous models, methods and tools have been proposed, which react on one or the more of these characteristics. All of these proposals have in common that in the lifespan of such systems different specifications are required and have to be han-



*G. Hommel and Sheng Huanye (eds.), Human Interaction with Machines,* 1-12.

assessment RM-ODP distinguishes the following five kinds of relationships around specifications:

1. The *conformance* of a realization to a specification, which means that the realization meets the requirements of the specification.
2. The *consistency* between two specifications, which means that there exists a realization which is conformant to both specifications.
3. The *validity* of a specification, which means that there exists a conformant realization of the specification.
4. The *compliance* of a specification $A$ with another specification $B$, which means that all requirements of the specification $B$ are met by any conformant realization of $A$.
5. The *equivalence* between two specifications $A$ and $B$, which means that a realization is conformant with $A$ iff it is conformant with $B$.

Other than in algebraic specification languages relationships which result from the composition of specifications are not addressed in RM-ODP. Composition is only treated on the level of modeling entities by notions of composition and decomposition of objects. Here a major difference between RM-ODP and algebraic specification languages is visible. RM-ODP does not prescribe any style of specification or specification language and, accordingly, does not support the concepts of compositionality of specifications. It simply leaves open whether or not the composition of objects is specified by a composition of specifications of these objects.

It turns out that the above five relationships have a close correspondence to traditional concepts in logic and can properly be addressed in terms of abstract logics as for example in [3]. If a realization is considered to be a model and a specification a set of sentences then conformance means satisfaction of sentences in a model. Consistency then is the existence of a model in which two or more specifications are satisfied, while validity is simply the existence of a model for a given specification. Compliance is the classical consequence relation between specifications and equivalence, accordingly, the classical concept of logical equivalence. Refinement then implies compliance and translation equivalence.

## 3 Observations on Algebras

Probably the first formal approach to reference points is given in the study of observational equivalence on algebraic specifications. In their paper [14] Sanella and Tarlecki ask the question what an observation on an algebra could be. Their question concerns the semantics of algebraic specifications

They argue in favor of a semantic concept which is an alternative to the well-known initial semantics approach (see for example [5]) where a specification is given meaning by the isomorphism class of initial algebras to that specification. In the alternative concept the meaning of a specification is defined to be a class of algebras which are behaviourally equivalent, thereby allowing realizations of a specification, which differ in their techniques but show the same behaviour (see [5] for further references).

The abstract model of development underlying the work in Sanellas and Tarleckis paper is as follows. Initially, requirements on the behaviour of a program are given. Then, a first specification is designed which satisfies these requirements. In a sequence of refinement steps further specifications are created which preserve the behaviour first specified. The final specification then is executable and may be regarded as a program which meets the initially given requirements. This development model may be seen as a very special case of the much more elaborate development model of RM-ODP. The theoretical considerations in [14] concern a mathematically defined concept of refinement and compatibility of refinement steps with horizontal structuring of specifications. Their concepts are stated in rather abstract terms and therefore apply for any institution.

In this framework a realization of a specification is an algebra, and an observation on an algebra is specified by a set of formulas (sentences in the given institution). Given the specification $\Phi$ of an observation and an algebra $A$ then an observation on $A$ with respect to $\Phi$ is the result of checking for all $\varphi$ in $\Phi$ the validity of $\varphi$ in $A$.

An instructive application of this concept of observation is to algebras, where certain sorts of data are marked as input sorts and other sorts of data are marked as output sorts. If observations are then specified by equations of terms, whose sorts belong to the set of output sorts, while the variables in these terms have sorts from the set of input sorts, an operational behaviour is specified which can be observed by providing input data and looking at the resulting output data. This example shows well the idea of reference points whose specification as a pair consists of the description of a location and the description of the expected behaviour at this location. Locations in this example are output sorts while the description of the expected behaviour at output sorts are the equations specifying the observation. The concept of Sanella and Tarlecki is not restricted to this algebraic case with equational specifications, but can generally be applied in the framework of any institution where a relationship between sentences and models is defined.

Using the concept of observation, the notion of observational equivalence is simply defined by saying that an algebra $A$ is observationally equivalent to an algebra $B$ with respect to a set of formulas $\Phi$ if for any

$\varphi \in \Phi$: $A \models \varphi$ iff $B \models \varphi$. The notion of *behavioural equivalence*, known in the literature, is just a special case of observational equivalence.

Yet another notion is derived from the concept of observation, the specification-building operation, which abstracts away from certain details of a specification by restricting its requirements to the specification of an observation. This specification-building operation may be called *observational abstraction*, in analogy to behavioural abstraction as defined in the specification language ASL (see [17] for references).

The concept of reference points, which is introduced in RM-ODP, very much follows the idea of observation as described in Sanella and Tarleckis paper. If reference points in ODP are indeed motivated from this concept of observation cannot be said for sure since RM-ODP does not provide any references to the literature, but it seems very likely that the authors of the reference model have been aware of the work on behavioural and observational abstraction.

## 4 Reference Points and Conformance Assessment in RM-ODP

The standardized Reference Model on Open Distributed Processing (RM-ODP) contains a framework for conformance assessment, in which reference points are used for conformance testing. Implicit in RM-ODP is the following abstract development model. Initially requirements are given which concern the use and operation of a system (in RM-ODP an open distributed system) in its environment. In the development process a number of specifications is created as outlined in Section 2 above. These specifications are produced in a sequence of development phases leading to the final realization. “After installation and commitioning, an instance of the ODP system then enters a phase of operational use, during which it should meet the needs expressed in the document specifying its requirements. The products ability to do so depends upon a number of practices during each of these development phases. Standards for providing ‘quality’ describe consistent sets of such practices along with the organization of people, documentation, and lifecycle stages to which they apply. Typically, some measurement of quality is made at each phase and changes are made if the quality is found to be lacking. Conformance assessment provides such a measure of quality, usually in the phase during which the implementation specification is realized. However, it may also be used in or have implication for other phases.” (see RM-ODP, Part 1: 9.1 [8]).

With several specifications across the different viewpoints of RM-ODP the so-called *multiple viewpoint problem* occurs: There is the danger of inconsistencies, since the relation between specifications from different viewpoints is in no case some kind of refinement. In order to avoid this problem it is necessary that the different specifications form a consistent set. RM-ODP therefore provides a framework for testing consistency and for verification that a realization is conformant to all specifications under the different viewpoints. This framework is composed from two steps: First, the establishment of *correspondences* between the concepts and architectural entities of the different specifications and second, the testing of behaviour at particularly chosen *conformance points*. For this purpose RM-ODP requires the determination of *conformance statements* for every specification. Such conformance statements consist in specified reference points and the specification of the behaviour that is expected at these points.

The role of reference points in conformance assessment is clearly stated in the following phrases of the standard: "When the conformance of the realization of an ODP specification is assessed using conformance testing, its behaviour is evaluated (by delivering stimuli and monitoring any resulting events) at specific (interaction) points. The points used are called *conformance points*, and they are usually chosen from a number of such points whose location is specified in the RM-ODP architecture. These potential conformance points are termed *reference points*." (see RM-ODP, Part 1: 9.3 [8]).

In the descriptive foundations of RM-ODP a reference point is seen as an organizational concept and defined to be an interaction point specified in an architecture for selection as a conformance point in a specification which is compliant with that architecture, while an interaction point is a location at which there exists a set of interfaces. Interfaces on the other hand are abstractions of the behaviour of an object and consist of a subset of interactions of that object together with a set of constraints on when they may occur.

In the computation and the engineering viewpoints of its architecture RM-ODP lists which reference points shall exist, and identifies for each viewpoint which obligations exist for an implementor who claims conformance of his realization to the different viewpoint specifications. His obligations mainly consist in the selection of reference points in the engineering viewpoint as conformance points and in the provision of specifications for the mapping between the abstract specifications and the implemented constructs. For this he will use the correspondences between the different specifications as they were prescribed by RM-ODP or defined by the architect. The concept of reference points and the framework of conformance assessment are described in the standard (see RM-ODP, Part 1: 9, Part 2: 8.6, 8.11, 10.6, 10.7 and 15, and in Part 3: 4.6, 5.3, 6.3, 7.3, 8.3, 9.3 and 10, [8]). An overall description of the involved concept can also be found in ([13], chapter 11).

## 5 Reference Points in the TINA Business Model

The Telecommunications Information Network Architecture Consortium (TINA-C) defined an RM-ODP based service architecture (see [18]). This reference model was meant to provide a general architectural standard for world wide operating telecoms to support interoperatibility and the provisioning of services. The TINA business model which is motivated from the enterprise concepts of RM-ODP, market analysis and intended architectural properties of the TINA standard, is an excellent example of a multi-domain model under the enterprise viewpoint.

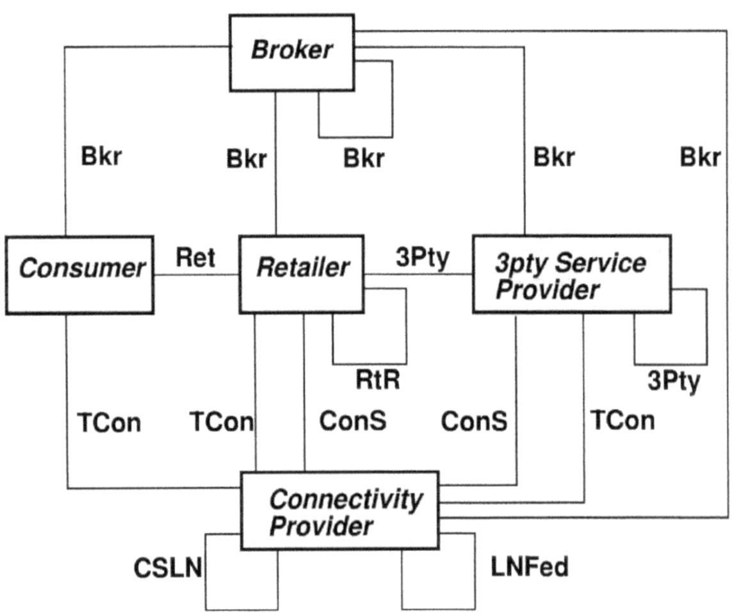


**Fig. 1.** The TINA initial business roles and business relationships.

It is highly generic and identifies five types of business administrative domains with the following raw interpretation: A consumer takes advantage of the services provided in a TINA system. A stakeholder in the retailer business role serves stakeholder in the consumer business role, a broker provides consumers with information that enables them to find other stakeholders and services in the TINA system, a third party provider supports retailers or other third-party providers with services, and a connectivity provider owns and manages a network for interoperability. The relationships between these business administrative domains are seen as reference points which specify conformance requirements by prescribing the interactions between stakeholders in these domains and interactions with the distributed processing environment (DPE) on which the TINA system is based.

Also TINA has an underlying model of development: Besides the specification and implementation of subsystems interoperability and compatibility between products from different venders are of major concern. A TINA system is assumed to be a world wide information networking system that is made up of a large number of interconnected computers, ranging from servers controlling huge databases to workstations and other user-side equipment. Since this equipment is owned by different stakeholders and may come from different venders there is the need for conformance assessment in a situation in which the system is emerging rather than being designed in a controlled way. Reference points in TINA are therefore prescriptive and directly constitute the concept of conformance in TINA: "An implementation of a TINA (sub-)system conforms to the Telecommunications Information Networking Architecture if one or more of the reference points defined in TINA are conformed to."

Two types of reference points are distinguished, inter-domain reference points and intra-domain reference points. The graph of the TINA business model names the inter-domain reference points in the marking of edges. Each reference point will consist of several viewpoint related specifications which are governed by a contract and can be seen as an aggregation of the specifications. Reference points thereby can be split into two parts, the part related to the application and the part related to the DPE supporting the application.

Since reference points in TINA tend to be large, as they consist of several specifications under each of the viewpoints, for testing purposes concepts have been developed which make the task of conformance testing easier. In their paper [15] Schieferdecker, Li and Rennoch introduce the concept of *reference points facets* which refine reference points in a way that specification, implementation and testing can be performed in a stepwise manner. The key idea of a reference point facet is that it is, so to say, context-independent in that it is self-contained and free from requirements from dependent operations.

## 6 Thoughts on a General Model for Consistency and Conformance

The discussion of the three approaches to reference points shows that there is not yet a common notion of reference point, but rather a collection of concepts which share a common intention and have similar features. The following are the differences:

- In the algebraic approach the idea of reference point is only implicit and there is no concept of location. Only in the briefly sketched example the sorts from the set of output sorts may be regarded as an abstract concept of locations. In the object-oriented approaches of RM-ODP and TINA locations are defined as interaction points and related to interfaces.
- RM-ODP does not require that the specification of a reference point also includes the specification of expected behaviour. Only if reference points are chosen to be conformance points, the behaviour specification is needed.
- TINA associates with a specification of a reference point a set of viewpoint specifications, which makes reference points complex entities at the systems level. RM-ODP instead requires that there exists a reference point at any interface of any computational object and that there is a reference point associated with every engineering object.

Besides these differences there are a number of commonalities which justify to compare the three approaches:

- In all three approaches reference points are not just interfaces which support abstraction by encapsulating the realization of certain behaviour, but they are perceived as means for the observation of behaviour through human machine interaction for inspection, model checking or testing.
- Reference points are in all three approaches embedded into a pragmatic context of systems development or validation. In each case, implicitly or explicitly, reference points concern matters of consistency and conformance.
- In each of the three approaches reference points are a matter of specification and are closely related to the systems architecture. Namely in RM-ODP and in TINA a framework for the architecture is prescribed and therefore the location of reference points for the possible observation of systems behaviour are identified. Since the algebraic approach

speaks of a program rather than a system there is no notion of architecture and reference points simply concern input / output behaviour.

The discussion of the three approaches shows increasing complexity in the according architectural theories, which results from the characteristics of today's open distributed systems. In this situation reference points seem to form a reasonable concept regard to consistency and conformance considerations. An abstract logic approach to specifications, requirements, behaviour, and observations, which is abstract enough to deal with the various forms of development processes and architectural frameworks, may be helpful in defining a generic model of consistency and conformance. In such a model consistency would have to be defined as the assertion that no inconsistencies can / will be observed in the systems use and operation. In such a model reference points would be needed in order to clarify the idea of observation and would therefore be associated with specifications and with conformance statements to allow the observation of expected behaviour. Reference points and conformance statements would then accompany the definition of functional behaviour related to the operational use of the component or subsystem specified. In such an abstract logic approach to specification there would be no structural assumptions on the relationship between specifications like particular types of correspondences, mappings or homomorphisms, but rather assertional assumptions as they are found in the concepts of conformance, compliance and consistency. From an abstract logic approach which follows these thoughts one would expect a rigorous conceptualization of conformance and consistency and, at the same time, a clarification and simplification of the highly involved explanations in RM-ODP and TINA.

## References


1. Sebastian Bab and Bernd Mahr. $\in_T$-integration of logics. In *Formal Methods in Software and Systems Modeling*, pages 204-219, 2005.
2. Marc Born, Eckhardt Holz, and Olaf Kath. *Softwareentwicklung mit UML 2*. Addison-Wesley Verlag München, 2004.
3. J.P. Cleave. *A Study of Logics*. Clarendon Press, 1991.
4. Hartmut Ehrig, Hans-Jörg Kreowski, Bernd Mahr, and Peter Padawitz. Algebraic implementation of abstract data types. *Theoretical Computer Science*, 20:209-263, 1982.
5. Hartmut Ehrig and Bernd Mahr. *Fundamentals of Algebraic Specification 1*. Springer Verlag Berlin, 1985.

6. Hartmut Ehrig and Bernd Mahr. *Fundamentals of Algebraic Specification 2.* Springer Verlag Berlin, 1990.
7. Hartmut Ehrig and Fernando Orejas. A generic component concept for integrated data type and process modeling techniques. Technical Report 2001/12, Technische Universität Berlin, 2001.
8. International Organization for Standardization. *Basic Reference Model of Open Distributed Processing*. ITU-T X.900 series and ISO/IEC 10746 series, 1995.
9. Joseph A. Goguen and Rod M. Burstall. Institutions: abstract model theory for specification and programming. *Journal of the ACM*, 39(1):95-146, 1992.
10. OMG.ORG: Object Management Group. Download of the MDA Specification. 2005.
11. OMG.ORG: Object Management Group. Download of the UML Specification. 2005.
12. T. Mossakowski. Foundations of heterogeneous specification. In M. Wirsing, D. Pattinson, and R. Hennicker, editors, *Recent Threads in Algebraic Development Techniques, 16th International Workshop, WADT 2002*, pages 359-375. Springer London, 2003.
13. Janis R. Putman. *Architecting with RM-ODP*. Prentice Hall PTR, 2000.
14. Donald Sannella and Andrzej Tarlecki. On observational equivalence and algebraic specification. In *TAPSOFT, Vol. 1*, pages 308-322, 1985.
15. Ina Schieferdecker, Man g Li, and Axel Rennoch. Formalization and testing of reference points facets.
16. A. Tarlecki. Towrds heterogeneous specification. In D. Gabbay and M. van Rijke, editors, *Proceedings 2nd International Workshop on Frontiers of Combining Systems, FroCoS'98*. Kluwer, 1998.
17. Martin Wirsing. Algebraic Specification. In *Handbook of Theoretical Computer Science, Volume B: Formal Models and Semantics (B)*, pages 675-788. 1990.
18. Martin Yates, Wataru Takita, Laurence Demoudem, Rickard Jansson, and Harm Mulder. Tina Business Model and Reference Points. 1997.

# Efficient Text Categorization Using a Min-Max Modular Support Vector Machine

Feng-Yao Liu[1], Kai-An Wang[1], Bao-Liang Lu[1*], Masao Utiyama [2], and Hitoshi Isahara [2]

[1] Department of Computer Science and Engineering
Shanghai Jiao Tong University
1954 Hua Shan Rd, Shanghai 200030, China
blu@cs.sjtu.edu.cn

[2] NICT Computational Linguistics Group
3-5 Hikaridai, Seika-cho, Soraku-gun, Kyoto, Japan
{mutiyama;isahara}@nict.go.jp

**Abstract.** The min-max modular support vector machine ($M^3$-SVM) has been proposed for solving large-scale and complex multiclass classification problems. In this paper, we apply the $M^3$-SVM to multilabel text categorization and introduce two task decomposition strategies into $M^3$-SVMs. A multilabel classification task can be split up into a set of two-class classification tasks. These two-class tasks are to discriminate class *C* from non-class *C*. If these two class tasks are still hard to be learned, we can further divide them into a set of two-class tasks as small as needed and fast training of SVMs on massive multilabel texts can be easily implemented in a massively parallel way. Furthermore, we proposed a new task decomposition strategy called hyperplane task decomposition to improve generalization performance. The experimental results indicate that the new method has better generalization performance than traditional SVMs and previous $M^3$-SVMs using random task decomposition, and is much faster than traditional SVMs.

## 1 Introduction

With the rapid growth of online information, text classification has become one of the key techniques for handling and organization of text data. Various pattern classification methods have been applied to text classification. Due to their powerful learning ability and good generation performance,

---

* To whom the respondence should be addressed. This research was partially supported by the National Natural Science Foundation of China via the grants NSFC 60375022 and NSFC 60473040, as well as Open Fund of Grid Computing Center, Shanghai Jiao Tong University.



*G. Hommel and Sheng Huanye (eds.), Human Interaction with Machines,* 13-21.

support vector machines (SVMs) [1] [2] have been successfully applied to various pattern classification problems. Joachims (1997) [3] and Yang (1999) [4] made experiments on the same text data set, respectively. Both experimental results showed that SVMs yield lower error rate than many other classification techniques, such as Naive Bayes and K-nearest neighbors. However, to train SVMs on large-scale problems is a time-consuming task, since their training time is at least quadratic to the number of training samples. Therefore, it is a hard work to learn a large-scale text data set using traditional SVMs.

On the other hand, Lu and Ito (1999) [5] proposed a min-max modular ($M^3$) neural network for solving large-scale and complex multiclass classification problems effortlessly and efficiently. And the network model has been applied to learning large-scale, real world multi-class problems such as part-of-speech tagging and classification of high-dimensional, single-trial electroencephalogram signals. Recently, Lu and his colleagues [6] have proposed a part-versus-part task decomposition method and a new modular SVM called min-max modular support vector machine ($M^3$-SVM), which was developed for solving large-scale multiclass problems.

In this paper, we will apply $M^3$-SVMs to multilabel text classification and adopt a new strategy of dividing a large-scale sample data set into many small sample data sets to try to investigate the influence of different task decomposition methods on the generalization performance and training time.

This paper is structured as follows. In Section 2, $M^3$-SVM is introduced briefly. In Section 3, several different task composition strategies are listed. Then in Section 4, we designed a set of experiments on a large-scale multilabel text classification. In Section 5, conclusions are outlined.

## 2 Min-Max Modular Support Vector Machine

Min-max modular support vector machine [6] is a method that divides a complex classification problem into many small independent two-class classification problems, learns these two-class problems in a parallel way, and then integrates these small SVMs according to two module combination rules, namely the minimization principle and the maximization principle [5].

For a two-class problem *T*, let $\chi^+$ denote the positive training data set belonging to a particular category ***C*** and $\chi^-$ denote the negative training data set not belonging to ***C***.

$$\chi^{+}=\left\{\left(x_i^{+},+1\right)\right\}_{i=1}^{l^{+}},\chi^{-}=\left\{\left(x_i^{-},-1\right)\right\}_{i=1}^{l^{-}} \quad (1)$$

where $x_i \in R^n$ is the input vector, and $l^+$ and $l^-$ are the total number of positive training data and negative training data of the two-class problem, respectively.

According to [6], $\chi^+$ and $\chi^-$ can be partitioned into $N^+$ and $N^-$ subsets respectively,

$$\chi_j^{+}=\left\{\left(x_i^{+j},+1\right)\right\}_{i=1}^{l_j^{+}}, \text{ for j=1, ... , } N^{+} \quad (2)$$

$$\chi_j^{-}=\left\{\left(x_i^{-j},-1\right)\right\}_{i=1}^{l_j^{-}}, \text{ for j=1, ... , } N^{-} \quad (3)$$

where $\bigcup_{j=1}^{N^+}\chi_j^+=\chi^+, 1\le N^+\le l^+$, and $\bigcup_{j=1}^{N^-}\chi_j^-=\chi^-, 1\le N^-\le l^-$.

After decomposing the training data set $\chi^+$ and $\chi^-$, the original two-class problem $T$ is divided into $N^+\times N^-$ relatively smaller and more balanced two-class subproblems $T^{(i,j)}$ as follows:

$$\left(T^{(i,j)}\right)^{+}=\chi_i^{+},\left(T^{(i,j)}\right)^{-}=\chi_j^{-} \quad (4)$$

where $\left(T^{(i,j)}\right)^{+}$ and $\left(T^{(i,j)}\right)^{-}$ denote the positive and negative data set of subproblem $T^{(i,j)}$ respectively.

In the learning phase, all the two-class subproblems are independent from each other and can be efficiently learned in a massively parallel way.

After training, the $N^+\times N^-$ smaller SVMs are integrated into a M$^3$-SVM with $N^+$ MIN units and one MAX unit according to two combination principles [5][6] as follows,

$$T^{i}(x)=\min_{j=1}^{N^-} T^{(i,j)}\left(x\right) \text{ and } T(x)=\max_{i=1}^{N^+} T^{i}\left(x\right) \quad (5)$$

where $T^{(i,j)}\left(x\right)$ denotes the transfer function of the trained SVM corresponding to the two-class subproblem $T^{(i,j)}$, and $T^{i}\left(x\right)$ denotes the transfer function of a combination of $N^-$ SVMs integrated by the MIN unit.

# 3 Two Types of Task Decomposition Strategies

Task decomposition is one of two key problems in the M$^3$-SVM. In this section, we introduce two types of task decomposition methods. One is the random decomposition strategy; and the other is the hyperplane decomposition strategy.

## 3.1 Random task decomposition strategy

The random task decomposition method is a simple and straightforward strategy. It means that we randomly pick up samples to form a new smaller and more balanced training data set. We refer to the M$^3$-SVM using random decomposition as M$^3$-SVM (R). Though the strategy can be implemented easily, it might lead to partial loss of statistical properties of original training data, so in some multiclass classification cases, it could result in a little decrease in the performance of text classification. Whereas, in the multilabel classifications, our M$^3$-SVM using random task decomposition method can also get better performance than SVMs.

## 3.2 Hyperplane decomposition strategy

An ideal decomposition strategy is the one without loss of the generalization performance. In order to achieve this goal, we hope to maintain the structural properties of the smaller data set as those of original data set after task decomposition. So we introduce a hyperplane strategy which divides original training set into smaller training sets using a series of hyperplanes. We regard the M$^3$-SVM using the hyperplane decomposition strategy as M$^3$-SVM (H). In some cases, we also let those training samples in these hyperplanes neighborhood simultaneously belong to two smaller and more balanced training set divided by hyperplanes which can be regarded as overlap of the small training set.

Suppose we divide the training data set of class $C_i$ into $N_i$ subsets. According to the above discussions, the M$^3$-SVM (H) method can be described as follows.

**Step 1.** Compute the distance between each training sample $x$ of class $C_i$ and hyperplane H: $a_1z_1 + a_2z_2, ..., +a_nz_n = 0$ as follows,

$$dist(x, H) = \frac{a_1x_1 + a_2x_2, ..., +a_nx_n}{\sqrt{a_1^2 + a_2^2, ..., +a_n^2}} \quad \textbf{(6)}$$

Where $x_i$ is the element of sample vector $x$.

**Step 2.** Sort the training data according to the value of $dist(x, H)$.

**Step 3.** Divide the ordered sequence of training data to $N_i$ parts equally to remain the size of sub-sets almost the same.

**Step 4.** Construct $M^3$-SVM according to Section 2.

From the above decomposition procedure, we can see that the hyper-plane task decomposition strategy can be easily implemented.

Now we construct an artificial data set illustrated in **Fig. 1.** Suppose that we want to divide the positive samples into two parts. The best way is that we take the hyperplane $X+Y=0$. Consequently, positive samples is partitioned into two comparatively smaller and more balanced data sets, and the structure of the data sets still remains unchanged after the combination according to the module combination principles of $M^3$-SVMs.

## 4 Experiments

In this section, we present experimental results for two mutillabel text classification problems to indicate that both the random task decomposition and the hyperplane task decomposition methods for $M^3$-SVM are effective. We use the Yomiuri News Corpus [7] and the revised edition of Reuters Corpus Volume I(RCV-v2) [8] for our study respectively. For the former, there are 913,118 documents in the full collections from the year 1996 to 2000. And we only use 274,030 documents which is 30 percent of the original collections as our training data set and 181,863 documents dated July-December of 2001 as our test set. There are totally 75 classes in this collection. For the RCV1-v2, we selected the four top classes, namely CCAT, ECAT, GCAT and MCAT, as the given classes. The training set we adopted here has a size of 23,149 samples, and the testing set 199,328 samples. The number of features for representing texts in the simulations is 5000 and 47,152 respectively. For our convenience, we regard the simulation using Yomiuri News collection as SIM(Y) and RCVI-v2 as SIM(R).

A multi-label task [9] can be split up into a set of two-class classification tasks. Each category is treated as a separate two-class classification problem. Such a two-class problem only answers the question of whether or not a document should be assigned to a particular category. Therefore our multi-label classification task can be converted into many two-class classification tasks. And each two-class classification task is decomposed into a series of two-class subproblems using different decomposition strategies. After training all two-class subproblems, we use min-max

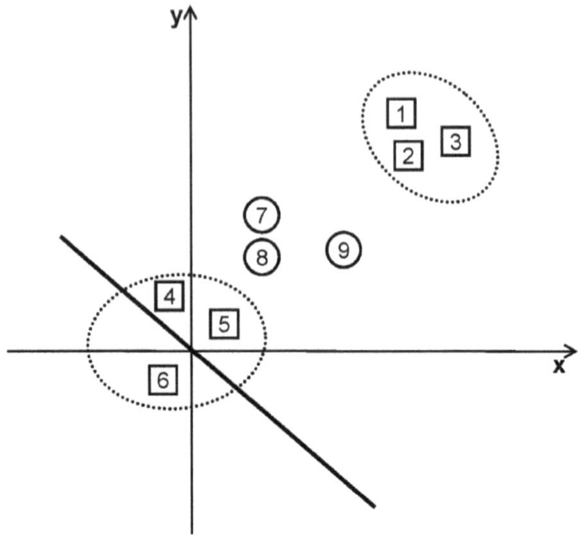


**Fig. 1.** An artificial data set in two-dimension space for making the explanation about effect if hyperplane task decomposition is used in $M^3$-SVMs. Here, square and circle denote positive and negative sample respectively. The solid line is so-called hyperplane.

combination strategies to integrate the trained individual SVMs into a $M^3$-SVM. All of the simulations were performed on an IBM p690 machine.

To compare the performance of $M^3$-SVM(H) with traditional SVM and $M^3$-SVM(R), the text classification in SIM(Y) is learned by traditional SVM and $M^3$-SVM(R) respectively, and in SIM(R), it is learned by traditional SVM, $M^3$-SVM(R) and $M^3$-SVM(H). In all the simulations, the kernel function of SVMs is linear kernel function and the hyperplane used in the task decomposition of SIM(R) is $z_1 + z_2, \ldots, +z_n = 0$.

In SIM(Y), we only list the results of top 10 classes and the parameter C we take is 8. The result is shown in **Table 1**. In SIM(R), we made four groups of experiments according to different tradeoffs C between training error and margin to the reciprocal of the average Euclidean norm of training examples. In each group, we take C as 0.5, 1, 2, 4, respectively, and also perform the simulations with 15% and 30% overlapping of the training samples. However, we only list the results on ECAT with C=0.5, since the results are comparatively best. The result is shown in **Table 2**. In this Table, 'o(0)', 'o(15)', 'o(30)' respectively stands for no overlap, 15% overlap, and 30% overlap of training data for $M^3$-SVM(H) method.

For more clear comparison, we organize experimental results of different methods in SIM(R) with all best generalization performance for each particular class into **Table 3**. From this table, we can see that $M^3$-SVM is a good choice for text classification problems.

**Table 1.** The simulation results for SIM(Y), where C=8

| Class Num | SVM | | | $M^3$-SVM | | |
|---|---|---|---|---|---|---|
| | $F_{1.0}$ | Time (h) | #SVM | $F_{1.0}$ | Serial (h) | Parallel (h) |
| 1 | 84.00 | 9.47 | 2 | 84.86 | 7.49 | 3.79 |
| 2 | 94.36 | 2.89 | 2 | 94.55 | 2.91 | 1.57 |
| 3 | 66.13 | 4.48 | 3 | 70.62 | 2.99 | 1.11 |
| 4 | 62.54 | 5.01 | 3 | 68.61 | 3.22 | 1.25 |
| 5 | 64.68 | 5.38 | 3 | 69.57 | 4.00 | 1.53 |
| 6 | 85.01 | 3.09 | 3 | 85.11 | 2.15 | 0.76 |
| 7 | 40.21 | 6.94 | 3 | 50.82 | 4.82 | 1.72 |
| 8 | 15.34 | 64.99 | 2 | 24.54 | 28.40 | 16.22 |
| 9 | 76.79 | 5.62 | 3 | 77.22 | 4.03 | 1.38 |
| 10 | 60.15 | 13.21 | 2 | 62.22 | 9.94 | 5.62 |

For evaluating the effectiveness of category assignments by classifiers to documents, we use the standard recall, precision and $F_{1.0}$ measure. Recall is defined to be the radio of the total number of correct assignments to correct assignments by the system. Precision is the ratio of the total number of the system's assignments to correct assignments by the system. The $F_{1.0}$ measure combines recall (r) and precision (p) with an equal weight in the following form:

$$F_{1.0} = \frac{2rp}{r+p} \tag{7}$$

From the experimental results shown in Tables 2 ans 3, we can draw the following conclusions:

**a.** Even though all of the individual SVMs were trained in serial, $M^3$-SVM including $M^3$-SVM(R) and $M^3$-SVM(H) is faster than traditional SVMs in our simulations. And with the increase of classifiers, $M^3$-SVMs need less and less training time.

**b.** In most cases, the generalization performance of $M^3$-SVM(R) is fluctuant. In some cases, $M^3$-SVM(R) has better generalization performance than traditional SVM and in other cases, the reverse occurs. Especially in SIM(Y), all the 10 classes we selected had a better performance by $M^3$-SVM(R) than by traditional SVMs. When the generalization performance is bad for some training set, for example for class ECAT, the generalization performance of $M^3$-SVM (R) can be raised by 4.6% at the best case and still by 1.9% at the worst case. And in SIM(Y), for class 7, the performance is raised from 40.21% to 50.82%.

**Table 2.** Results on ECAT in SIM(R), where C=0.5

| Method | | #SVM | CPU time (s.) | | Performance | | |
|---|---|---|---|---|---|---|---|
| | | | parallel | serial | P | R | $F_{1.0}$ |
| SVM | | 1 | 607 | 607 | 92.7 | 64.1 | 75.8 |
| $M^3$-SVM(R) | | 3 | 186 | 531 | 84.7 | 74.7 | 79.4 |
| | | 7 | 53 | 347 | 73.8 | 82.5 | 77.9 |
| | | 20 | 21 | 365 | 78.5 | 78.3 | 78.4 |
| | | 26 | 16 | 365 | 74.5 | 81.1 | 77.7 |
| $M^3$-SVM(H) | O(0) | 3 | 234 | 461 | 87.9 | 71.0 | 78.6 |
| | | 7 | 66 | 307 | 80.4 | 78.9 | 79.6 |
| | | 20 | 34 | 300 | 82.6 | 77.5 | 80.0 |
| | | 26 | 26 | 310 | 79.0 | 80.6 | 79.8 |
| | O(15) | 3 | 233 | 519 | 89.1 | 70.0 | 78.4 |
| | | 7 | 72 | 350 | 82.0 | 77.5 | 79.7 |
| | | 20 | 36 | 363 | 83.8 | 76.9 | 80.2 |
| | | 26 | 28 | 373 | 80.2 | 79.9 | 80.0 |
| | O(30) | 3 | 239 | 590 | 89.9 | 68.9 | 78.0 |
| | | 7 | 85 | 428 | 83.7 | 76.3 | 79.8 |
| | | 20 | 42 | 451 | 84.1 | 76.7 | 80.3 |
| | | 26 | 34 | 473 | 81.2 | 79.9 | 80.4 |

**c.** In all the cases in SIM(R), $M^3$-SVM(H) show better generalization performance than traditional SVMs and $M^3$-SVM(R). And with the increasing number of classifiers, $M^3$-SVM(H) has better and better generalization performance.

**d.** In all the cases, $M^3$-SVM(R) need less training time than traditional SVMs and $M^3$-SVM(H).

## 5 Conclusions

We have presented two task decomposition strategies for multilabel text classification. The advantages of the proposed methods over traditional SVMs are its parallelism and scalability. And compared with $M^3$-SVM(R), $M^3$-SVM(H) has better generalization performance. When overlap ratio of training set become high, $M^3$-SVM(H) have better generalization performance. With the increase of the number of classifiers, the performance of $M^3$-SVM(H) could reach maximum. A future work is to search for breakpoint between overlap ratio and the number of classifiers and analysis the effectiveness of the hyperplane decomposition strategy theoretically.

**Table 3.** Comparison of generalization performance and training time among methods in SIM(R), where C=0.5 and CAT stands for category

| CAT | Method | #SVM | CPU time (s.) | | Performance (%) | | |
|---|---|---|---|---|---|---|---|
| | | | parallel | serial | P | R | $F_{1.0}$ |
| CCAT | SVM | 1 | 898 | 898 | 94.9 | 91.3 | 93.0 |
| | M$^3$-SVM(R) | 4 | 290 | 989 | 94.9 | 90.7 | 92.8 |
| | M$^3$-SVM(H) O(0) | 30 | 47 | 563 | 93.5 | 92.7 | 93.1 |
| | M$^3$-SVM(H) O(30) | 30 | 60 | 894 | 94.0 | 92.6 | **93.3** |
| ECAT | SVM | 1 | 607 | 607 | 92.7 | 64.1 | 75.8 |
| | M$^3$-SVM(R) | 3 | 186 | 531 | 84.7 | 74.7 | 79.4 |
| | M$^3$-SVM(H) O(0) | 20 | 34 | 300 | 82.6 | 77.5 | 80.8 |
| | M$^3$-SVM(H) O(30) | 26 | 34 | 473 | 81.2 | 79.7 | **80.4** |
| GCAT | SVM | 1 | 785 | 785 | 85.5 | 89.1 | 92.2 |
| | M$^3$-SVM(R) | 8 | 126 | 675 | 92.7 | 91.9 | 92.3 |
| | M$^3$-SVM(H) O(0) | 8 | 141 | 552 | 93.5 | 91.0 | 92.2 |
| | M$^3$-SVM(H) O(30) | 8 | 182 | 937 | 93.9 | 90.9 | **92.4** |
| MCAT | SVM | 1 | 636 | 636 | 94.5 | 87.2 | 90.7 |
| | M$^3$-SVM(R) | 4 | 184 | 663 | 87.9 | 93.7 | 90.7 |
| | M$^3$-SVM(H) O(0) | 24 | 31 | 355 | 91.2 | 90.8 | 91.0 |
| | M$^3$-SVM(H) O(30) | 24 | 39 | 540 | 91.7 | 90.8 | **91.3** |

## References


1. C. Cortes and V. N. Vapnik, "Support-vector network", Machine Learning, Vol. 20 (1995) 273-297
2. V. N. Vapnik, Statistical Learning Theory, John Wiley and Sons, New York, 1998
3. Thorsten Joachims, "Text categorization with support vector machine: Learning with many relevant features", Technical report, University of Dortmund, Computer Science Department, 1997
4. Yiming Yang and Xin Liu, "A re-examination of text categorization methods", In: Proceedings of the ACM SIGIR Conference on Research and Development in Information Retrieval, 1999
5. B. L. Lu and M. Ito, "Task Decomposition and Module Combination Based on Class Relations: a Modular Neural Network for Pattern Classification", IEEE Transactions on Neural Networks, 1999, Vol.10, pp.1244-1256
6. B. L. Lu, K. A. Wang, M. Utiyama, H. Isahara, "A part-versus-part method for massively parallel training of support vector machines", In: Proceedings of IJCNN'04, Budapast, July25-29 (2004) 735-740
7. M. Utiyama and H. Ichapire. Experiments with a new boosting algorithm. In Proceedings of 13th International Conference on Machine Learning, 1996, 148-156
8. David D. Lewis, Yiming Yang, Tony G. Rose, Fan Li, "RCV1: a new benchmark collection for text categorization research", Journal of Machine Learning Research 5(2004) 361-397
9. Thorsten Joachims, Learning to classify text using support vector machine:method, theory, and algorithms. Kluwer Academic Publishers, 2002.

The first Bluetooth specification (version 1.0) was published in July 1999, but due the manifold ongoing development activities it is fair to say that Bluetooth is still a moving target. Moreover, our experience (as well as that of others) shows that there are still interoperability problems between devices of different brands. It is difficult to get a representative picture of how widespread such problems are, but the mailing list archives of the official Blue-tooth implementation of the Linux kernel (BlueZ [7]) indicate that they are not negligible.

If two devices fail to interoperate properly it is – viewed from a practical standpoint – not helpful to debate *who is to blame*; rather, one should find out *what can be done* about it. It is our thesis that a self-healer can often bridge such gaps of misunderstanding.

However, it is not our intention to develop a new protocol stack implementation which closely integrates self-healing techniques. Instead we aim at a framework that allows us to augment *existing implementations* with self-healing capabilities.

In Sect. 2 we look at other work in the self-healing area. Section 3 presents our proposal for a self-healing Bluetooth protocol stack architecture and Sect. 4 describes our prototype implementation. Section 5 gives a self-healing example and Sect. 6 concludes and outlines directions of further work.

## 2 Related Work

To the best of our knowledge no work has been published about self-healing for wireless communication problems caused by erroneous implementations. Most work in the wireless realm concerned with failure avoidance and/or recovery focuses either on transmission problems and their consequences or on higher-level network issues. Examples of the former are radio interference between Wi-Fi and Bluetooth and how its effects can be alleviated [11], or upper-layer throughput dips due to lost link layer packets and how an adaptive packet selection scheme can improve the situation [3]. On the network level the topic of automatic scatternet construction gets much attention, e.g. [4, 8], where the healing aspect is the topological restructuring performed after devices join and leave the scatternet.

Some inspiration can be drawn from projects that deal with self-healing but are unrelated to wireless communications, e.g. the UC Berkeley/Stanford cooperation *Recovery-Oriented Computing* (ROC), or parts of Carnegie Mel-lon University's *Architecture Based Languages and Environments*(ABLE). The ROC project places emphasis on failure recovery instead of failure avoidance. Interesting ROC concepts are recursive restartability, where recovery time is improved by hierarchically organized partial restarts, and crash-only software, where recovery operations are so tightly integrated into the system as to make them indistinguishable from normal operation. These concepts augment each other leading to the notion of microreboots [2]. The ABLE project aims to lay foundations for specification and analysis of software architectures and architectural styles. With regard to self-repairing systems, external mechanisms are proposed that interpret monitored parameters in reference to an architectural run-time model [6]. Repair operations modify the architectural model according to style constraints and propagate changes back into the real system.

Because both projects focus on Internet services, inherent assumptions of their techniques are not fulfilled in our context. But some principles like strong component isolation and externalization of healing facilities can be recognized in our approach.

## 3 A Self-healing Architecture for Protocol Stacks

There is no established design for self-healing for Bluetooth stacks. Therefore we had to develop a new self-healing architecture. Even though we will instantiate our architecture only for the case of Bluetooth (in Sect. 3.2), our design goals are not limited to Bluetooth stacks, but are also suitable for other layered protocol stacks. Therefore we formulate them more generally in Sect. 3.1.

### 3.1 Design Goals for a Protocol Stack Self-healing Architecture

The design goals of our self-healing architecture can be summarized as follows:

**Architectural protocol stack compatibility.** The self-healing architecture needs to fit to the existing protocol design and should not fight its natural data flow. Often the protocol structure follows a layered architecture and reflects different abstraction levels that the self-healer can take advantage of.

**Minimally invasive add-on.** Since implementations of Bluetooth stacks are inherently complex, the self-healing capability must not increase this complexity. Moreover, the implementors will rarely possess expertise with self-healing techniques. Therefore the self-healer has to be an add-on which affects the original program as little as possible.

**API.** To ensure minimal invasiveness we only base on a well defined and small API that the implementors of a Bluetooth stack have to provide in order to enable self-healing for their program.

**Agent-based structure.** The self-healer must work independently from the main stack. Therefore it must in particular not run in the same execution thread(s). As a matter of fact, it should ideally act as a distinguished *agent*or as a *collection of agents*. We come back to this topic in Sect. 4.

**Flexible self-healing strategy.** Hardcoding of specific solutions can quickly lead to a confusing multitude of ad-hoc hacks with unintended side effects. A well-defined *strategy*of recognition and healing gives more structure to the whole process, hopefully not at the expense of *flexibility*. This will be elaborated in Sect. 3.3.

**Updatable healing knowledge.** It is not realistic to expect a self-healer in an embedded device to adapt itself automatically to wrong behaviors of other devices and changed firmware. Therefore it must be possible to *update the healing knowledge*in order to keep the self-healer up to date.

### 3.2 A Bluetooth Self-healing Architecture

With the aforementioned goals in mind we developed the architecture depicted in Fig. 1. It contains the following main components.

**Bluetooth Controller.** As a hardware device the Bluetooth Controller is not directly amenable to self-healing, but it may be possible to work around some behavioral bugs using the standardized HCI interface (see Sect. 5 for an example) or vendor-specific commands. However, if the controller fails severely and there is no second device available to act as a substitution, self-healing has obviously reached its limits.

**Bluetooth Host Stack.** The Bluetooth Host Stack contains the usual components and layers (HCI, L2CAP, SDP, etc.). But the connections between the layers – depicted by the dotted arrows – are different from classical stack implementations: The data flow actually takes a detour along the solidly drawn arrows, giving the self-healer the opportunity to interfere with the original stack on a layer-by-layer basis.

This modified stack design constitutes the *self-healing API* that has to be supported by Bluetooth host stacks, if they shall enable self-healing.

The API allows the healer to absorb and insert arbitrary numbers of packets, and thus to initiate and control complex communication sequences. The interaction between host stack and self-healer is based on *data entities specified by the protocol* itself, abstracting from internals of the chosen host stack.

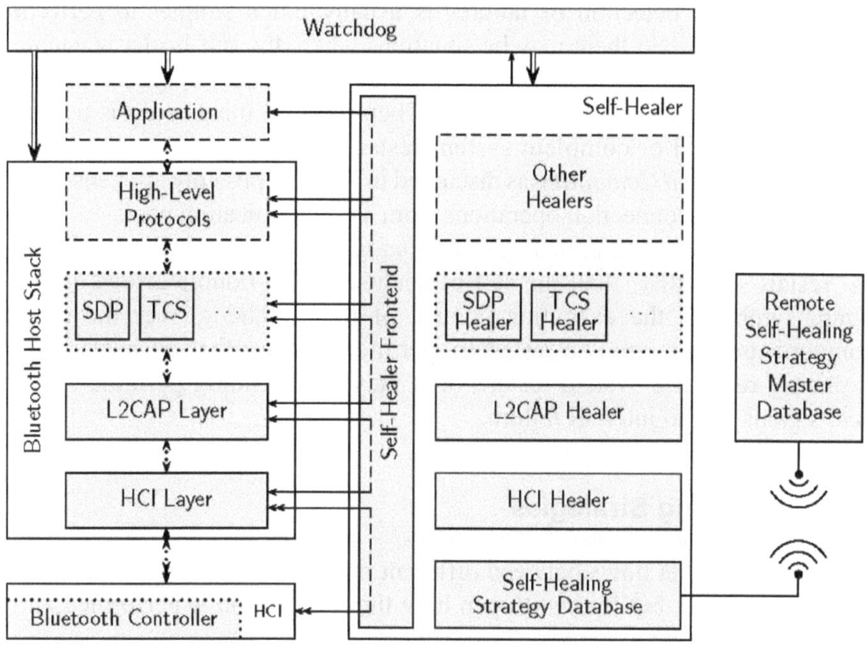


**Fig.1.** Proposed Bluetooth self-healing architecture.

**Application.** The application utilizes the host stack to provide the end-user functionality. It is specific to the concrete deployment scenario. Its connection to the self-healer needs further investigation.

**Self-healer.** In order to obtain a clear conceptual design, the architecture of the healer mirrors that of the original stack. Hence, there are sub-healers like the HCI healer, the L2CAP healer, etc.

To keep the requested API simple, we have introduced a front-end which shields the inner workings of the healer from the stack and acts as communication peer. It decides which of the healer's components (if any) are invoked for the processing of received packets.

The overall working of the healer is guided by the contents of the *self-healing strategy database*. Each entry in the database is supposed to determine the conditions under which the healer becomes active and what it does. This will be elaborated in more details in Sect. 3.3.

**Remote Self-healing Strategy Master Database.** From time to time the local database is reconciled with the external vendor-provided master database and brought up to date with regard to self-healing knowledge. To this end a suitable communication path is needed, e.g. data transmission via a mobile phone.

**Watchdog.** The detection of failures is usually much simpler to perform than their healing. So there may be situations where the self-healer is unable to keep the system going, but it knows that the system has failed or can predict that the system is about to fail. Then the best thing to do is usually to cause a partial or complete system restart (following the principles of *Recovery-Oriented Computing* as discussed in Sect. 2), possibly after ensuring quasi-normal disconnection operations from a communication peer.

An independent component, the *watchdog*, can act much more reliably as "restart specialist" than the self-healer itself. The double arrows in the figure symbolize the execution control of the watchdog over the other components. There are two restart trigger mechanisms: the self-healer may explicitly request a system restart, or it may stop sending periodic heartbeat signals due to internal failure.

## 3.3 Self-healing Strategies

Monitoring of data flows between different protocol layers is a prerequisite to detect failures, but leaves it open how the recognition is performed and how failures can be repaired.

### Recognition of Failures

One obvious way to recognize failures is to look for errors/exceptions propagating between the layers. However, that is in general not feasible for various reasons. For example, the generation of an error response by the signaling layer may cause an internal state change that is difficult to revoke thereafter. Even more intricate is the problem that many errors are not caused by a single message (command packet, event packet, etc.), but by an inconsistent *sequence* of messages. Therefore a message that triggers an error may not be the true reason for a failure. Consequently, there will be no repair for this message per se and the healing process may depend on the history of preceding messages. There are even situations that do not lead to errors at all but nevertheless need healing to ensure proper operation (as can be seen in our concrete case study in Sect. 5).

Therefore we need a *proactive strategy* for error recognition. We plan to build on results achieved in the MWatch project [9] where a language for temporal trace specifications was developed, roughly similar to regular expressions extended with timing constraints, e.g. admissible delays.

By a *trace* we mean a sequence of messages with timing properties. A *critical trace* is a trace that leads to an erroneous situation, which the healer should recognize and repair. The MWatch tool takes a set of such trace specifications and constructs an automaton for recognizing them. Once constructed, this automaton is efficient enough for real-time use.

We conceive the following overall scenario. Over time, reports about errors encountered in the field are used to build up a body of knowledge in a master database in form of a collection of critical traces (cf. Fig. 1). The download of the corresponding automaton to the local database then enables the self-healer to recognize critical traces and to take associated actions.

## Repair of Failures

There are degrees of healability.

- In the simplest case, the healer may replace an offending message by another message or simply change message parameters.
- In more complex cases, one may replace a whole sequence of messages by another sequence, thus turning a critical trace into a normal trace.

Unfortunately, the second approach is di ûcult in practice due to the request-response scheme used by most Bluetooth protocols: the second quest will not be issued until the first response arrives – and this cannot happen, when the request is queued instead of sent.

Thus, in most cases the healer can only act after a (complete) critical trace has been encountered (which may or may not end with an error). In-general the healing action consists of the invocation of a specialized code fragment taken from the database. This is flexible, but lacks structure and may impose too heavy storage demands for an embedded device.

A possible solution could be predefined template-based traces (TBT) that can be "replayed" as healing actions. Such a TBT is a compact description of parameterized packets to be sent and responses to be expected and decomposed. The packet templates are filled with parameters extracted from earlier responses and with data from the critical trace.

If a critical trace is recognized for which no healing action is known, a smooth system restart via the watchdog is initiated as a last resort. The same technique is used if an error message is encountered that is not registered among the critical traces. Since in this case the situation is new and unknown, it is sensible to report the circumstances to the master database for further analysis. This contribution to the body of assembled knowledge can then be used by human experts to devise an appropriate strategy for the future.

# 4 A Self-healing Prototype

In order to experiment with our ideas depicted in the previous sections, we have implemented a prototype in Java.

## 4.1 The Host Stack

We have chosen to use the JavaBluetooth stack [10] as the target for our self-healing prototype. This choice was primarily motivated by the easy accessibility for experimentation, but since the stack originally uses UART dongles, we had to find a way to attach modern USB dongles.

We perform our development in the Linux environment, so we decided to leverage the kernel-integrated Bluetooth stack BlueZ [7] for that purpose. BlueZ offers support for L2CAP, SDP, RFCOMM, BNEP, etc. However, we ignore all these high-level features and utilize only BlueZ' low-level HCI access facilities to talk to the USB dongle. The message exchange with the dongle is completely controlled by our prototype.

## 4.2 Instantiation of the Bluetooth Self-healing Architecture

A major issue in the implementation is the underlying "agent" model. Our perception is heavily influenced by the process concept of the language Erlang [5]. This programming language offers built-in support for concurrency, distribution, fault tolerance, updating of running systems, and is suited for soft real-time applications. It has been deployed in several large telecommunication systems. Erlang processes are very lightweight, allowing for extremely fast context-switching. They run independently from each other and exchange messages via a message-queue communication system.

Java's concurrency support is based on threads and encourages a rather different programming style. While objects encapsulate methods and state, they do not encapsulate threads which are an orthogonal concept. Many threads can visit methods of the same object at the same time. Information exchange between threads is done by changing state of "shared" objects. Thus synchronization of threads is of paramount importance in order to prevent state corruption. In contrast to threads, agents cannot corrupt each others state since they run in different address spaces.

From the object-oriented perspective, we roughly consider an agent to be a set of objects which are associated with exactly one thread. Figure 2 shows how the self-healer front-end is embedded into two queues, one for the messages from the self-healer to the host stack, and the other for the messages from the host stack to the self-healer. The right queue is also used for healer-internal timer-controlled event handling. Messages are enqueued by the sending thread and dequeued by the receiving thread. Queue operations are quick and independent of the rest of the system. Even if the communication peer has locked up, an agent will not hang when accessing a queue.

The current prototype is limited to the HCI layer and does not yet support the trace-based recognition mechanism discussed in Sect. 3.3. It works on the basis of a simpler scheme of dynamic lookup tables for HCI command and event handlers.

## 5 A Self-healing Example

Bluetooth devices use a specific inquiry procedure to determine the hardware addresses of potential communication partners in their vicinity. We have run across a device – call it device X – with a strange firmware bug:

- A *normally completed* inquiry effects an unacceptable slowdown of the subsequent communication, when more than one device is in reach.
- However, if inquiry is *aborted* before normal completion, things work as expected.

This example illustrates that sometimes one cannot wait for problems to manifest themselves but needs to act beforehand (proactive).

A simpliÞed form of the self-healing procedure is depicted in Fig. 3. When the host stack initiates an inquiry, some event handlers and timers are set up. The reception of an Inquiry Result event causes the self-healer to consider the time elapsed since inquiry was started. If normal inquiry completion is impending, the inquiry procedure is immediately canceled. Otherwise abortion is scheduled for an appropriate point in future time to give other devices the chance to respond. When inquiry canceling completes, an Inquiry Complete event is faked by the self-healer, so that the host stack has the impression of normal inquiry completion. Since we learned that inquiry terminates prematurely if a command is sent to our controller during inquiry, we take special care to not forward host stack commands to the controller until inquiry is over. Note that we do not halt the host stack, but enter a selective dequeuing mode to delay command packet processing. Halting the host stack would not work because we need its lowest layer to talk to the controller.

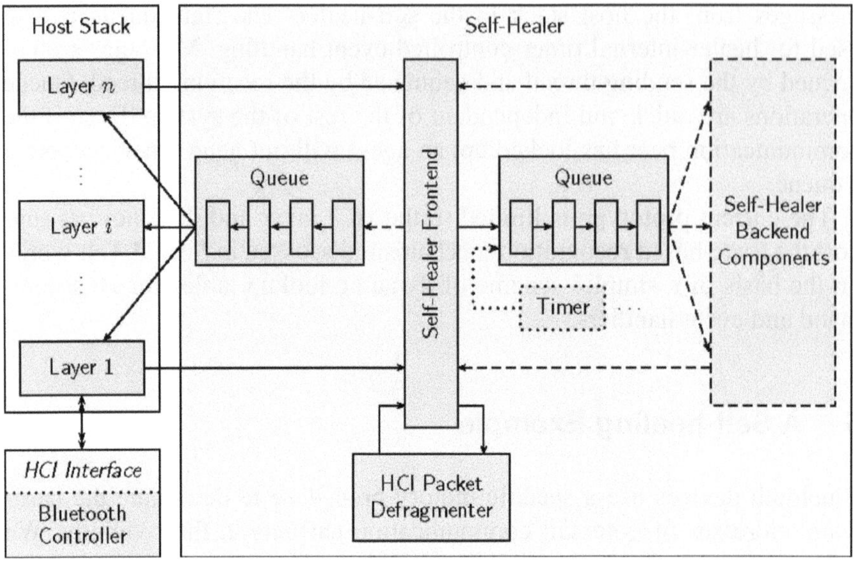


**Fig. 2.** The self-healer front-end mediates between host stack and self-healer. Solid arrows denote calls made by host stack threads, dashed arrows correspond to the self-healer thread. The HCI Packet Defragmenter is only needed because of a Java-Bluetooth bug and may be unnecessary for other host stacks

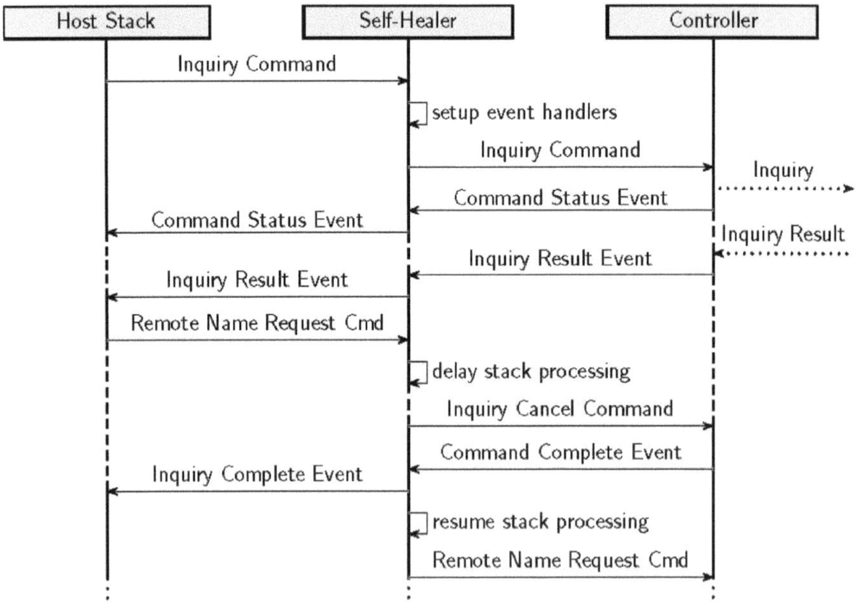


**Fig. 3.** Self-healing procedure for inquiry (see text).

## 6 Conclusions and Future Work

Our experiment provides a number of key insights into the problems of self-healing for embedded systems such as the Bluetooth stack.

- Self-healing can be feasible, even when no redundant hardware or ample computation power is available.
- The self-healing architecture should mirror the architecture of the original system as closely as possible.
- Self-healing should and can be based on minimally invasive techniques. This can be achieved in the form of well-designed APIs.
- A mandatory feature is the disentanglement provided by an advanced agent model in the style of Erlang. The Java thread system is – in its plain form – unsuited for such purposes.

Among the main areas where we intend to focus future work are the prototype, which is supposed to expand its realm to higher layers, and the topic of self-healing strategies, where various details need to be matched to the Blue-tooth environment. We are confident that we can benefit in this context from earlier work on trace specification and pattern recognition. However, Blue-tooth is quite unfriendly to self-healing, so close inspection of problems and chances is required – there is no silver bullet.

## Acknowledgement.

We thank or colleagues from the UBB group at the TU Berlin for valuable comments. Particular thanks go to Wieland Holfelder and Kal Mos from Daimler-Chrysler RTNA for drawing our attention to the Bluetooth example and for many elucidating discussions on the topic.



## References

[1] Bluetooth. Official membership site. http://www.bluetooth.org
[2] Candea G, Kawamoto S, Fujiki Y, Friedman G, Fox A (2004) Microreboot – a technique for cheap recovery. Proc 6th USENIX/ACM Symp on Operating Systems Design and Implementation (OSDI'04)
[3] Chen LJ, Kapoor R, Sanadidi MY, Gerla M (2004) Enhancing Bluetooth

TCP throughput via link layer packet adaptation. Proc IEEE Int Conf on Communications, vol 7, pp 4012–4016
[4] Cuomo F, Melodia T, Akyildiz IF (2004) Distributed self-healing and variable topology optimization algorithms for QoS provisioning in scatternets. IEEE Journal on Selected Areas in Communications, 22(7):1220–1236
[5] Erlang Programming Language. http://www.erlang.org
[6] Garlan D, Cheng SW, Schmerl B (2003) Increasing system dependability through architecture-based self-repair. In: de Lemos R, Gacek C, Romanovsky A (eds) Architecting Dependable Systems. Springer
[7] Holtmann M, Krasnyansky M, et al. BlueZ. http://www.bluez.org
[8] Jayanna D, Z'aruba GV (2005) A dynamic and distributed scatternet formation protocol for real-life Bluetooth scatternets. Proc 38th Hawaii Int Conf on System Sciences (HICSS'05)
[9] Lepper M (2004) An Algorithm for the Real-Time Evaluation of Temporal Trace Specifications. PhD thesis, Technische Universität Berlin, Germany
[10] Lorenz C. JavaBluetooth stack. http://www.javabluetooth.org
[11] Ophir L, Bitran Y, Sherman I (2004) Wi-Fi (IEEE 802.11) and Bluetooth coexistence: Issues and solutions. Proc 15th IEEE Int Symp on Personal, Indoor and Mobile Radio Communications (PIMRC'04), vol 2, pp 847–852

# From Keyframing to Motion Capture

Yan Gao, Lizhuang Ma, Xiaomao Wu, Zhihua Chen

Department of Computer Science and Engineering
Shanghai Jiao Tong University, Shanghai 200030, China
gaoyan@sjtu.edu.cn ma-lz@cs.sjtu.edu.cn
wuxiaomao@sjtu.edu.cn zhihua@sjtu.edu.cn

**Abstract.** A survey is given for human motion synthesis techniques from the viewpoint of human computer interaction. The focus is on an overview based on the interaction of animation systems between human and computer. To design a good interactive system, we concern the underlying structure of such a system more than the concrete interface that the system will adopt. Therefore, we first analyze the tasks of animators for designing a human motion synthesis system. After reviewing the evolution of human motion synthesis, we try to reveal how such evolution improves interaction between human and computer. Finally, we demonstrate our motion synthesis system and give some suggestions for future research directions.

## 1 Instruction

Many synthetic scenes involve virtual humans, from special effects in the film industry to virtual reality and video games. However, despite great advances in recent years, creating effective tools for synthesis of realistic human motion remains an open problem in computer animation. This is particularly due to two reasons. One is because animating virtual humans range from various motion types such as walking, running and diving etc. Another is because people have proven to be adept at discerning the subtleties of human movement and identifying inaccuracies.

An ideal character animation system should be able to synthesize believable motion, and, at the same time, provide sufficient control to the generated motion. "*Believable*" here has the same meaning proposed by Bates in his term "*believable agent*" [1] – *a believable character is one who seems lifelike, whose actions make sense, who allows you to suspend disbelief.* In computer animation, believability is generally provided by animator through his fine control to the synthesized motion. So to interact



*G. Hommel and Sheng Huanye (eds.), Human Interaction with Machines,* 35-42.

with the virtual human is in a way equal to control it to perform user-defined tasks.

Nowadays, research in HCI has been spectacularly successful and has fundamentally changed computing. In spite of great progress in HCI, there is widespread misconception about HCI, which equal it to interface layout concerning menu arrangement, color selection and so on. In fact, for a good user interface design, there are a lot of thing to do far more than interface layout. As Lewis [2] mentioned, "*No matter how well they are crafted, the interface will be a failure if the underlying system doesn't do what the user needs, in a way that the user finds appropriate. In other words, the system has to match the users' tasks.*" That is to say, while the interface is important, we must also look beyond it to the underlying structure of the worlds we want to model.

So to design a good animation system for human computer interaction, the most important thing is not to adopt which types of input devices or interfaces but how to transform the interaction information hidden behind the interfaces. In other words, we concern control contents more than control forms. In this paper, we review the evolution of human motion synthesis and try to reveal the tendency of the change of interaction information between animator and computer. Moreover, we evaluate the influence on interface design caused by the different motion synthesis methods.

The remainder of this paper is organized as follows: In section 2 we discuss the general human computer interaction model of human motion synthesis tools and section 3 presents current motion synthesis techniques. Section 4 introduces some new interaction methods applied in motion synthesis field. In section 5, we show the experimental results of our synthesis system. Finally, we conclude with a brief discussion in Section 6.

## 2 Human Computer Interaction Model

A computer animation system can be regarded as a system that maps input to output. Its output is the result motion clips and its input varies according to system design. The whole interaction process can be shown as follows:

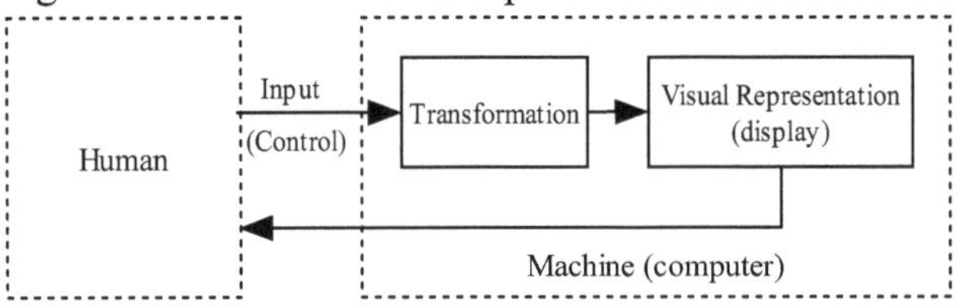


**Fig. 1.** Human computer interaction.

From the above model, we conclude that to design a good animation system is somewhat equal to construct a good mapping from input to output. For human motion synthesis, this becomes a challenging task because human motion involves much more degrees of freedom than the usual computer input devices can provide. Furthermore, we argue that to design a good synthesis, the most important thing is to make a deep understanding to user's demands. For a human animation system, we express its design tasks as "believable motion" and "minimal user assistance" while its interaction tasks as "efficiency" and "low input dimensions".

Here, "believable motion" is explained in the above section while "minimal user assistance" means that system can generate desired motion with animator's assistance as few as possible. "Efficiency" emphasizes that system can interact with animators in real-time while "low input dimensions" favors a few input parameters to be controlled by user.

## 3 Evolution of Human Motion Synthesis

The basic approaches to human motion synthesis for articulated figures include: keyframing, procedural methods and motion capture. Though these techniques emerge in sequence, we cannot conclude that one technique excels the others absolutely.

### 3.1 Keyfarming

The most common approach used by animators is keyframing. The animators position the body for important frames (called keyframes). The computer generates the intermediate frames (called the inbetweens) by interpolating the values at adjacent keyframes. Keyframing is a natural extension of the most traditional animation methods: draw each frame by hand.

The earlier animation systems are all keyframing systems. The process of interaction between human and computer for keyfarming systems is shown in figure 2.

Keyframe systems provide users with absolute control over the creative process. However, it is a notoriously time consuming task to synthesize complex movements such as walking and running. This intrinsic limitation

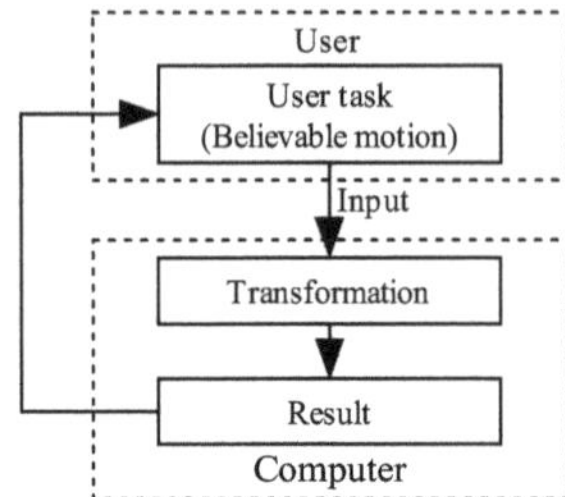


**Fig. 2.** HCI model for keyframing.

makes keyframing systems hard to meet "efficiency" and "low input dimensions" demands.

## 3.2 Procedural Methods

In order to reduce the burden on the animator, there has been a large amount of research to develop techniques based on procedural animation. Procedural methods explicitly describe how movements are generated, and use computation to propagate changes in high-level parameters to changes in low-level parameters. The process of interaction between human and computer for procedural animation systems is shown in figure 3.

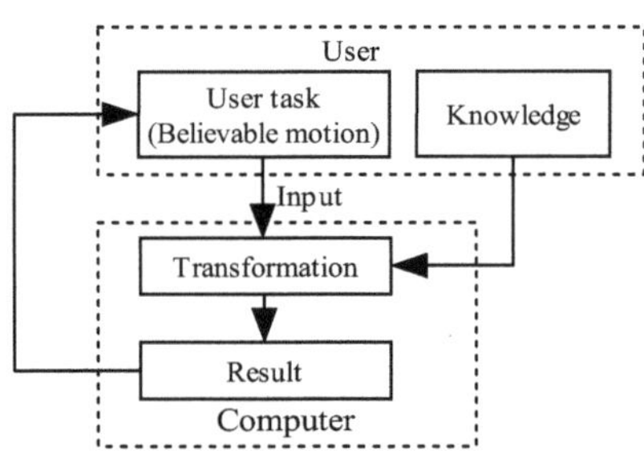


**Fig. 3.** HCI model for physical simulation.

From figure 3, we can see that, relative to keyframing, knowledge is added in the interaction model of procedural animation systems. Knowledge can be any of biomechanical knowledge, physical laws and so on. Such knowledge can help animators to generate believable motion clips at the cost of the reduction of user controls. However, current procedural methods do not scale up to the complexity of characters one would like to animate. As a result, they fail in the "efficiency" demand and it is hard to meet the "believability" task.

### 3.3 Motion Capture

All the previous approaches deal with biomechanical, mechanical or empirical knowledge on human motion. In contrast, motion capture method works directly on data, which are recorded by motion capture systems. In general, the term *motion capture* refers to any method for obtaining data that describes the motion of a human or animal.

Though motion capture can provide all the detail and nuance of live character motion, the process of acquiring motion data is labor intensive and costly. Moreover, the captured data are closely related to the specific capture environment so that they cannot be directly used in new environments. In order to make motion capture widely available, it is desirable to develop tools that make it easy to reuse and adapt existing motion data. The model of interaction between human and computer for motion capture techniques is shown in figure 4.

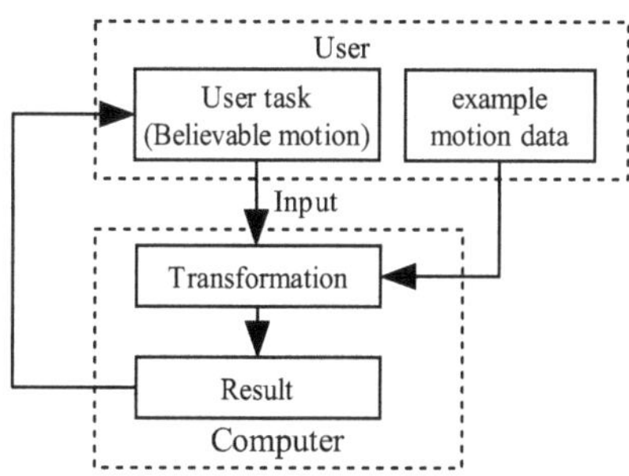


**Fig. 4.** HCI model for motion capture.

Motion capture is the most promising technique for human motion synthesis. It can satisfy the "believable motion" and "minimal user assistance" demands, which make it competent for realistic, motion synthesis task. However, if motion data are modified beyond allowed "region", the result motion is unacceptable. Moreover, motion capture methods fail to satisfy the "low input dimensions" demand.

## 4 New Interaction Techniques

As the progress of HCI, many interaction means, other than conventional mouse-based techniques (menus, scroll bars, buttons, etc.), have been developed to improve the interaction between human and computer. These new interfaces include sketching, script, gesture, speech and video etc. They help to overcome the "curse of dimensions" problem: the character's motion is high dimensional and most of the available input devices are not.

## 4.1 Script Based Systems

Script language, which aims to provide high-level control, is often used as a good interactive means between human and computer. For example, many commercially available animation systems such as *Jack* [3] incorporated the specification of low level motion commands, even though these commands do not extend easily to controlling the motion of articulated figures.

## 4.2 Sketch Based Systems

Sketching provides an intuitive way for users to express themselves between human and computer. It is especially effective for novice users. Thorne and his colleagues [4] presented a sketch-based technique for creating character motion. In this system, the user not only sketches the character to be animated, but also sketches motion curves that are segmented and mapped to a parameterized set of output motion primitives.

## 4.3 Foot-Ground Force Based Systems

Some researchers used foot-ground contact information as input for motion synthesis. For example, Yin and Pai [5] presented a *FootSee* System which used a foot pressure sensor pad to interactively control avatars for video games, virtual reality, and animation. During the training phase, his system captures full body motions with a motion capture system, as well as the corresponding foot-ground pressure distributions with a pressure sensor pad, into a database. At run time, animation is created by searching the database for the measured foot-ground interactions.

## 4.4 Gesture Based Systems

Gestures are a promising approach to human computer interaction because they often allow several parameters to be controlled simultaneously in an intuitive fashion. Zhao [6] developed a framework for the procedural generation of expressive and natural-looking gestures for computerized communicating agents. In his system, input gestures are parameterized based on the theory of *Laban Movement Analysis* (LMA) and its *Efforts* and *Shape* components. New gestures are then synthesized based on predefined key pose and time information plus *Effort* and *Shape* qualities.

### 4.5 Video Based Systems

Vision-based interface, due to the higher dimensional nature of the input, gives the most control over the details of the avatar's motion. Lee [7] constructed a graph and generated motion via three user interfaces: a list of choices, a sketch-based interface, and a live video feed. In his vision-based interface, the user acts out the desired motion in front of a camera. Visual features are extracted from video and used to determine avatar motion.

## 5 Our Experiment Results

We have developed a human motion synthesis platform by combining *Inverse Kinematics* with motion capture techniques. The user modifies end-effector's positions of the character through mouse. The system computes the other degrees of freedom of character automatically and generates whole motion through interpolation. Our graphic user interface is very friendly and convenient for manipulation. The results are shown in figure 5.

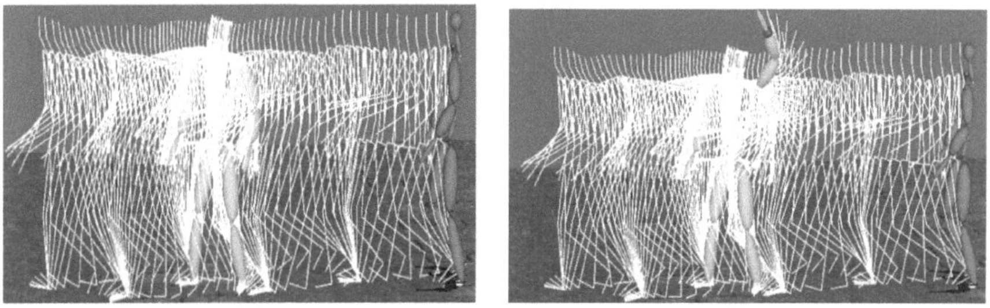

(a) (b)

**Fig. 5.** (a) Original motion: walk and look behind midway. (b) The left motion is edited to look behind and wave midway.

## 6 Conclusions

Creating realistic human motion by computer has long been a challenging task and needs animator's tedious assistant work. To reduce animator's task, many motion synthesis techniques have been developed during the past decades. In this paper, instead of introducing these techniques, we focus on analyzing interaction between human and computer of these animation systems. Therefore our paper acts a bridge between computer animation and human computer interaction.

The interaction of motion synthesis tools suffers from the problem of "curse of dimensions" – the dimensions of human's degrees of freedom are high while input devices have low dimensions. To state how this problem influence the system's design; we first review the history of human motion synthesis techniques and give a comparison to three main techniques: key-framing, procedural methods and motion capture techniques. We hope other people can benefit from our paper to create more efficient means for human motion synthesis.

New interfaces also help to alleviate the "curse of dimension" problem greatly. These new interfaces such as video-based interface, gesture-based interface etc. can provide more useful information to user than the usual mouse and keyboard interface. We believe that by using these new interfaces and hybrid synthesis methods, new interaction tools, which need user provide his desired high-level tasks, can be developed.

## Acknowledgements

This material is based on work supported by the Microsoft Foundation No. Project-2004-Image-01.

## References

[1] Bates J (1994) The role of emotion in believable agents. In: *Communications of the ACM*, vol 37(7): p122-125.

[2] Lewis C, Rieman J Task-Centered user interface design (online book). available at http://hcibib.org/tcuid

[3] Badler N, Phillips C, Webber B (1993) Simulating Humans: Computer Graphics, Animation, and Control. *Oxford University Press*.

[4] Thorne M, Burke D, Panne VD (2004) Motion doodles: An interface for sketching character motion. In: *Proceedings of SIGGRAPH 2004*, p424-431.

[5] Yin K, Pai DK (2003) FootSee: an Interactive Animation System. In: *Proceedings of the Eurographics/SIGGRAPH Symposium on Computer Animation*, San Diego.

[6] Zhao L (2002) Synthesis and acquisition of Laban Movement Analysis qualitative parameters for communicative gestures. *PHD thesis*, University of Pennsylvania.

[7] Lee J, Chai J, Reitsma PSA, Hodgins JK, Pollard NS (2002) Interactive control of avatars animated with human motion data. In: *ACM Transactions on Graphics*, 21(3):491-500.

# Multi-Modal Machine Attention: Sound Localization and Visual-Auditory Signal Synchronization

Liqing Zhang

Department of Computer Science and Engineering, Shanghai Jiao Tong University, Shanghai 200030, China

**Abstract:** This paper discusses machine attention based on multimodal information integration and synchronization. The aim is to introduce a computational model for robust multimodal attention mechanism in artificial systems. First, we introduce a new blind source separation algorithm based on temporal structures and high order statistics of the source signals. Combining the bandpass filter and blind source separation, we present a new sound localization approach to find the sound position in a complicated environment. Furthermore, we introduce a visual-auditory signal synchronization approach to focus visual attention in the artificial system to the interesting object that synchronizes the auditory and visual signals.

## 1 Motivation

Machine attention is the first step towards machine intelligence. Generally it is not easy to well define the concept of machine attention. The most important function of selective visual attention is to direct our gaze rapidly towards objects of interest in the visual environment. A two-component framework for attentional deployment has recently emerged, suggesting that subjects selectively direct attention to objects in a scene using both bottom-up, image-based saliency cues and top-down, task-dependent cues.

Usually, human perception and attention have multimodal character, whereas hearing and vision seem to be complementary strategies. Human attention and eye movements tightly interplay, posing computational challenges with respect to the coordinate system used to control visual attention from auditory sound localization. Insights from neurobiological experiments provide a framework for a computational and neurobiological understanding of auditory-visual multimodal attention [4], [6].

The field of machine attention has been studied for quite long time and a large number of models and applications exists - biologically motivated approaches or technical solutions of specific application problems. How-



*G. Hommel and Sheng Huanye (eds.), Human Interaction with Machines,* 43-50.

ever, multimodal approaches based on auditory-visual system are still relatively seldom. Thus, modeling early auditory-visual integration is promising significant advantage in the machine attention [5], [7].

When microphones receive more than one sound/speech, the conventional sound localization approach does not work well. In such a case, we need apply first blind separation method to extract an interesting sound from microphone recordings, which are convolutive mixtures of different sounds in the scene. Blind source separation or independent component analysis (ICA) has been considered as a fundamental data analysis tool in the fields of signal processing and neural network [1], [3], [8]. We will use a new ICA approach to localize sound position in the complicated environment.

## 2 Independent Residual Analysis

In this section, we formulate the blind separation of temporally correlated sources in the framework of independent residual analysis. The objective function is derived from the mutual information of the residual signals. Learning algorithms for the demixing model and temporal structures are also provided.

### 2.1 Problem Formulation

Assume that $s_i$, $i = 1,2,\cdots,n$ are $n$ mutually stochastically independent source signals, of which each is temporally correlated and zero mean. Suppose that source $s_i(k)$ is modeled by a stationary AR model

$$s_i(k) = \sum_{p=1}^{N} a_p^i s_i(k-p) + \varepsilon_i(k) \tag{1}$$

where $N$ is the degree of the AR model and $\varepsilon_i(k)$ is referred to as the residual, which is zero mean, and independent and identically distributed (i.i.d.). For the sake of simplicity, we use the notation

$$A_i(z) = 1 - \sum_{p=1}^{N} a_p^i z^{-p}. \tag{2}$$

where $z$ is the z-transform variable.

If source signals are known, the coefficients of the forward linear predictor $A_i(z)$ can be obtained by minimizing the prediction error in the least squares sense. The linear filter $A_i(z)$ is assumed to minimum phase, that is the zeros of $A_i(z)$ are located in the interior of the unit circle.

$$A(z) = diag(A_1(z), \cdots, A_n(z)) = \sum_{p=0}^{N} A_p z^{-p} \tag{3}$$

where $A_0 = I$ is the identity matrix and $A_p = -diag(a_p^1, \cdots, a_p^n), p \geq 1$ is a diagonal matrix. We consider the case when observed signals are instantaneous mixtures of the source signals. Let $x(k) = (x_1(k), \cdots, x_n(k))^T$ be the set of linearly mixed signals,

$$x_i(k) = \sum_{j=1}^{n} H_{ij} s_j(k) \tag{4}$$

Here, $H = (H_{ij})$ is an $n \times n$ unknown nonsingular mixing matrix. For the sake of simplicity, we assume that the number of observations is exactly the same as that of the source signals.

Blind source separation is to find a linear transform which transforms the sensor signals into mutually independent components, which are used to recover source signals. Let $W$ be an $n \times n$ nonsingular matrix which transforms the observed signals $x(k)$ to

$$y(k) = Wx(k), \tag{5}$$

where $x(k) = Hx(k)$. In order to derive a cost function for blind separation, we suggest using the mutual information of residual signals of output signals as a criterion. The conventional methods usually employ the mutual information of the output signals as a criterion. The purpose of exploring the residual signals is to take the temporal structures of the source signals into consideration in the learning algorithm.

When the temporal structure $A(z)$ and the demixing matrix $W$ are not well-estimated, the residual signals

$$r(k) = A(z)Wx(k), \tag{6}$$

are not mutually independent. Therefore, it provides us a new criterion for training the demixing model and temporal structures to make the residuals $r(k) = A(z)Wx(k)$ mutually independent. To this end, we employ the mutual information rate of random variables as a criterion. We intend to find the demixing matrix $W$ and the temporal filters $A(z)$ such that the output signals $r(k)$ are spatially mutually independent and temporally identically independently distributed.

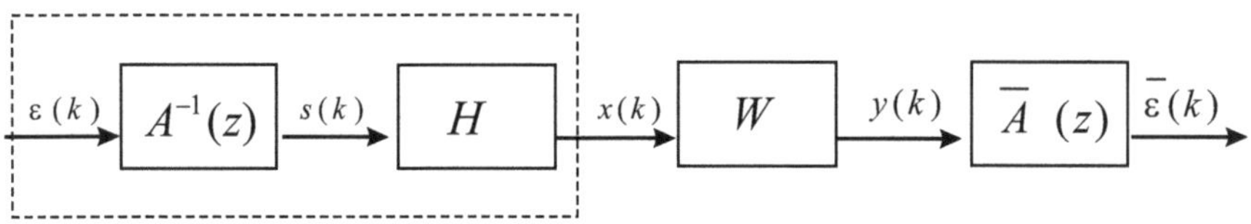


**Fig. 1.** General Framework of Independent Residual Analysis

Now we introduce the mutual information rate $I(r)$ between a set of stochastic processes $r_1, \cdots, r_n$ as

$$I(r) = H(r) - \sum_{i=1}^{n} H(r_i), \tag{7}$$

which measures the mutual independency between the stochastic processes $r_1, \cdots, r_n$, where $H(r_i)$ is the entropy of random variable $r_i$. Minimizing the mutual information with respect to the demixing matrix and temporal structure provides us a new approach to developing the learning algorithm for the demixing matrix and the temporal structures of the sources.

## 2.2 Learning Algorithm for Demixing Matrix

In this section, we derive a learning algorithm based on the gradient descent approach for the demixing matrix. Because the filter in (2) is assumed to be minimum-phase, the cost function for on-line statistical learning can be simplified from the mutual information (7) as follows

$$L(W, A(z)) = -\log(|\det(W)|) - \sum_{i=1}^{n} \log q_i(r_i) \tag{8}$$

where $q_i(r_i)$ is the probability density function of the residual signal $r_i$, and $r(k) = A(z)Wx(k)$. By minimizing the cost function (8), we derive the gradient descent algorithm for training $W$ [1],

$$\Delta W = \eta\left\{ I - \sum_{p=0}^{N} [A_p \varphi(r) y^T(k-p)] \right\} W \tag{9}$$

where $\varphi(r) = (\varphi_1(r_1), \cdots, \varphi_n(r_n))^T$ is the vector of activation functions, defined by $\varphi_i(r_i) = -\frac{q_i'(r_i)}{q_i(r_i)}$. There are two unknowns in the algorithm, the probability density function $\varphi(r)$ and the temporal filter $A(z)$. These two unknowns can be estimated via maximum likelihood, such that the equilibriums of the cost function are stable.

## 3 Sound Localization

In this section, we present a new framework of sound localization in a complicated environment. First, a binaural model is introduced to localize the sound of interest in the ideal case. Then sub-band sound localization is presented based on the blind separation of sub-band signals and binaural model.

### 3.1 Binaural Model

In contrast to visual perception, hearing starts with one–dimensional, temporal signals, whose phasing and spectrum are essential for the localization. To localize a sound of interest, the auditory system utilizes acoustic effects caused by a distance variation between the sound source and the two ears. The most significant acoustics effects are intensity differences and time delays in the two ears. In [2] a comprehensive study of sound localization based on different types of binaural and monaural information is presented. The estimator precision in the horizontal plane depends on the relation of azimuth angle variation and interaural time differences (ITDs). The source localization tasks generally could be solved by calculating ITDs and the detailed functional and structural description of the ITDs. Refer to [2] for detailed information. The interaural time differences can be estimated via the interaural cross correlations

$$\Theta(\tau) = \left\langle x_{left}(t), x_{right}(t-\tau) \right\rangle \tag{10}$$

where $x_{left}(t)$, $x_{right}(t)$ are the signals received from the left and right ears, respectively, $t$ is the time parameter, and $\tau$ is the time difference between two ears. The time difference can be estimated by maximizing the cross correlations function $\Theta(\tau)$.

### 3.2 Subband-Source Localization

In general, the source localization approach does not work well when the environment is noisy or the number of sound sources is larger than one. In order to improve the sound localization performance, more complicated model and algorithms should be applied. In this paper, we present a new framework of sound localization by using the sub-band information of sources and blind separation approach. In a complicated environment, the model for the recorded signals can be described a convolutive model as follows

$$x(k) = \sum_{p=0}^{L} H_p s(k-p) \tag{11}$$

where $s(k) = (s_1(k), \cdots, s_n(k))^T$ denotes the vector of $n$ source signals which are assumed to be mutually stochastically independent, $H_p$ is an $n \times n$ -dimensional matrix of mixing coefficients at time-lag $p$, called the impulse response at time $p$, and $x(k) = (x_1(k), \cdots, x_n(k))^T$ is an $n$-dimensional vector of sensor signals, which are convolutive mixture of source signals.

Blind deconvolution problem is to recover original sources only from measurements by using a demixing model. Generally, we carry out the blind deconvolution with another multichannel LTI system of the form

$$y(k) = \sum_{p=0}^{L} W_p x(k-p), \tag{12}$$

where $y(k) = (y_1(k), \cdots, y_n(k))^T$ is an $n$ -dimensional vector of the outputs and $W_p$ is an $n \times n$ -dimensional coefficient matrix at time lag $p$, which are the parameters determined during training. However, the blind deconvolution problem itself is usually much more difficult than sound localization problem. Therefore, we will not attempt to extract one sound of interest, but use some sub-band information to perform the sound localization talks.

Applying the windowed Fourier transform, the relationship between observations and sources is approximated by

$$\hat{x}(\omega,t_s) = \hat{H}(\omega)\hat{s}(\omega,t_s) \tag{13}$$

where $\hat{H}(\omega)$ is the Fourier transform of filter matrix $H(z)$, and $\hat{s}(\omega,t_s)$ is the windowed Fourier transform of source signals $s(k)$. For fixed frequency $\omega$ in Eq. (13), $\hat{x}(\omega,t_s)$ is regarded as an instantaneous mixture of complex-valued time series $\hat{s}(\omega,t_s)$. Therefore, the convolutive mixture problem is divided into quite simple subproblems of non-convolutive mixtures at frequency $\omega$. Note that in the above representation, the filter matrix is parameterized in the frequency domain and matrix $\hat{H}(\omega) = \sum_{\tau} H_\tau e^{j\omega\tau}$ for one frequency can be estimated independently of other frequencies.

## 4 Multi-modal Synchronization

Neurobiological study shows Superior Colliculus (SC) area in the brain, provide evidence for a merging of the sensor modalities and the forming of multisensory spatial maps [4], [5]. Our computational model is inspired by findings about the Inferior Colliculus in the auditory pathway and the visual and multimodal sections of the Superior Colliculus. Accordingly it includes: a) an auditory map, based on interaural time delays, b) a visual map, based on spatio-temporal intensity difference and c) a bimodal map where multisensory integration is performed. Multimodal map can be described a modified Amari-type neural field as follows

$$\begin{aligned}\tau\frac{d}{dt}z(r,t) &= -z(r,t) + W_A x_A(r,t) + W_V x_V(r,t) \\ &- c_i \int y(z(r,t))dr + c_e \int \Phi(r-r')y(z(r',t))dr',\end{aligned} \tag{14}$$

where the state field $z(r,t)$ at position $r$ depends on three components, the multimodal inputs and global inhibition and lateral feedback from neighboring positions $r'$. The outputs of neurons are given by the sigmoidal function $y(z(r,t)) = (1+\exp(-\sigma.z(r,t)))^{-1}$. The bimodal inputs $x_A(r,t)$ and $x_V(r,t)$ are the internal representations of the auditory and visual neural information respectively. The auditory-visual synchronization is defined by the spatial correlations between the two internal neural signals

$$\Omega(v) = \int \langle x_A(r,t), x_V(v-r,t) \rangle_r \, dt \,. \tag{15}$$

If the maximum value of the spatial correlation $\Omega(v)$ exceeds given threshold, we call the auditory and visual signals are synchronized. The synchronized point now is considered as the object of interest, which the artificial system should focus on it. The motor system will control the object of interest to the center of the visual scene. After the machine vision is directed to the object of interest, we can use the vision information, such as the lip movement, to improve the speech recognition performance. We can also explore both auditory and visual multimodal information to identify the person's identification. Due to the limited space, we should leave the detailed discussion on the visual-motor control to another full paper.

## Acknowledgment

The project was supported by the National Natural Science Foundation of China under grant 60375015 and the Scientific Research Foundation for the Returned Overseas Chinese Scholars, State Education Ministry, China.

## References

[1] S. Amari. Estimating function of independent component analysis for temporally correlated signals. *Neural Computation*, 12(9):2083-2107, 2000.
[2] J. Blauert. *Spatial Hearing.* MIT Press, 1996.
[3] A Cichocki and S. Amari. *Adaptive Blind Signal and Image Processing*. John Wiley, Chichester, UK, 2003.
[4] Daniel C., et al, The influence of visual and auditory receptive field organization on multisensory integration, Exp Brain Res 139:303-310, 2001.
[5] MH. Giard, F. Peronnet, Auditory-visual integration during multimodal object recognition in humans, *Journal of Cognitive Neuroscience*, 1999.
[6] J. Lazzaro and C. Mead, A silicon model of auditory localization. *Neural Computation*, 1(1):41-70, 1989.
[7] C. Schauer, T. Zahn, P. Paschke, HM. Gross, Binaural Sound Localization in an Artificial Neural Network, *Proc. IEEE ICASSP*, 2000, pp865-868.
[8] L. Zhang, A. Cichocki, and S. Amari. Self-adaptive blind source separation based on activation function adaptation. *IEEE Transactions on Neural Networks*, 15(2):233-244, 2004.

# A note on the Berlin Brain-Computer Interface

K.-R. Müller[1,2], M. Krauledat[1,2], G. Dornhege[1], S. Jähnichen[1], G. Curio[1], and B. Blankertz[1]*

[1] Fraunhofer FIRST.IDA, Kekuléstr. 7, 12 489 Berlin, Germany [2] University of Potsdam, August-Bebel-Str. 89, 14 482 Potsdam, Germany [3] Technical University Berlin, Str. d. 17. Juni 135, 10 623 Berlin, Germany [4] Dept. of Neurology, Campus Benjamin Franklin, Charité University Medicine Berlin, Hindenburgdamm 30, 12 203 Berlin, Germany

**Summary.** This paper discusses machine learning methods and their application to Brain-Computer Interfacing. A particular focus is placed on linear classification methods which can be applied in the BCI context. Finally, we provide an overview on the Berlin-Brain Computer Interface (BBCI).

## 1 Introduction

Brain-Computer Interfacing is an interesting, active and highly interdisciplinary research topic ([2, 3, 4, 5]) at the interface between medicine, psychology, neurology, rehabilitation engineering, man-machine interaction, machine learning and signal processing. A BCI could, e.g., allow a paralyzed patient to convey her/his intentions to a computer application. From the perspective of man-machine interaction research, the communication channel from a healthy human's brain to a computer has not yet been subject to intensive exploration, however it has potential, e.g., to speed up reaction times, cf. [6] or to supply a better understanding of a human operator's mental states.

Classical BCI technology has been mainly relying on the adaptability of the human brain to biofeedback, i.e., a subject learns the mental states required to be understood by the machines, an endeavour that can take months until it reliably works [7, 8].

The Berlin Brain-Computer Interface (BBCI) pursues another objective in this respect, i.e., to impose the main load of the learning task on the 'learning machine', which also holds the potential of adapting to specific tasks and changing environments given that suitable machine learning (e.g. [9]) and adaptive signal processing (e.g. [10]) algorithms are used. Short

* The studies were partly supported by the *Bundesministerium für Bildung und Forschung* (BMBF), FKZ 01IBB02A and FKZ 01IBB02B, by the *Deutsche Forschungsgemeinschaft* (DFG), FOR 375/B1 and the PASCAL Network of Excellence, EU # 506778. This paper is based on excerpts of [1].



*G. Hommel and Sheng Huanye (eds.), Human Interaction with Machines,* 51-60.

training times, however, imply the challenge that only few data samples are available for learning to characterize the individual brain states to be distinguished. In particular when dealing with few samples of data (trials of the training session) in a high-dimensional feature space (multi-channel EEG, typically several features per channel), overfitting needs to be avoided. It is in this high dimensional – small sample statistics scenario where modern machine learning can prove its strength.

The present paper introduces basic concepts of linear classification (for a discussion of nonlinear methods in the context of BCI, see [11, 9, 1, 12, 13]). Finally, we briefly describe our BBCI activities where some of the discussed machine learning ideas come to an application and conclude.

## 2 Linear methods for classification

In BCI research it is very common to use linear classifiers, but although linear classification already uses a very simple model, things *can* still go terribly wrong if the underlying assumptions do not hold, e.g. in the presence of outliers or strong noise which are situations very typically encountered in BCI data analysis. We will discuss these pitfalls and point out ways around them.

Let us first fix the notation and introduce the linear hyperplane classification model upon which we will rely mostly in the following (cf. Fig. 1, see e.g. [14]). In a BCI set-up we measure $k = 1 \ldots K$ samples $\mathbf{x}_k$,where $\mathbf{x}$ are some appropriate feature vectors in $n$ dimensional space. In the training data we have a class label, e.g. $y_k \in \{-1,+1\}$ for each sample point $\mathbf{x}_k$. To obtain a linear hyperplane classifier

$$\mathbf{y} = \text{sign}\,(\mathbf{w}^{\mathrm{T}}\mathbf{x} + b) \tag{1}$$

we need to estimate the normal vector of the hyperplane $\mathbf{w}$ and a threshold $b$ from the training data by some optimization technique [14]. On unseen data $\mathbf{x}$, i.e. in a BCI feedback session we compute the projection of the new data sample onto the direction of the normal $\mathbf{w}$ via Eq.(1), thus determining what class label $\mathbf{y}$ should be given to $\mathbf{x}$ according to our linear model.

### 2.1 Optimal linear classification: large margins versus Fisher's discriminant

Linear methods assume a linear separability of the data. We will see in the following that the optimal separating hyperplane from last section maximizes the *minimal* margin (minmax). In contrast, Fisher's discriminant maximizes the *average* margin, i.e., the margin between the class means.

**Large margin classification**

For linearly separable data there is a vast number of possibilities to determine $(\mathbf{w}, b)$, that all classify correctly on the training set, however that vary in quality on the unseen data (test set). An advantage of the simple hyperplane classifier (in canonical form cf. [15]) is that literature (see e.g. [14, 15]) tells us how to select the *optimal* classifier $\mathbf{w}$ for unseen data: it is the classifier with the largest margin $\rho = 1/\|\mathbf{w}\|_2^2$, i.e. of minimal (euclidean) norm $\|\mathbf{w}\|_2$ [15] (see also Fig. 1). Linear Support Vector Machines (SVMs) realize the large margin by determining the normal vector $\mathbf{w}$ according to

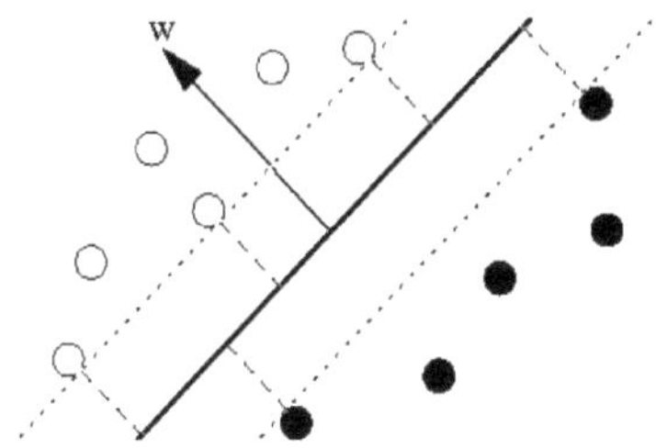


**Fig. 1.** Linear classifier and margins: A linear classifier is defined by a hyperplane's normal vector $\mathbf{w}$ and an offset $b$, i.e. the decision boundary is $\{\mathbf{x} \mid \mathbf{w}^T \mathbf{x} + b = 0\}$ (thick line). Each of the two halfspaces defined by this hyperplane corresponds to one class, i.e. $f(\mathbf{x}) = \mathrm{sign}(\mathbf{w}^T\mathbf{x} + b)$. The *margin* of a linear classifier is the minimal distance of any training point to the hyperplane. In this case it is the distance between the dotted lines and the thick line. From [9].

$$\min_{\mathbf{w},b,\xi} \; 1/2 \, \|\mathbf{w}\|_2^2 + C/K \, \|\xi\|_1 \quad \text{subject to} \tag{2}$$

$$y_k(\mathbf{w}^T \mathbf{x}_k + b) \geq 1 - \xi_k \text{ and}$$

$$\xi_k \geq 0 \text{ for } k = 1, \ldots, K,$$

where $\|\cdot\|_1$ denotes the $\ell_1$-norm: $\|\xi\|_1 = \sum |\xi_k|$. Here the elements of vector $\xi$ are slack variables and parameter $C$ controls the size of the margin vs. the complexity of the separation. While the user has not to care about the slack variables, it is essential to select an appropriate value for the free parameter $C$ for each specific data set. The process of choosing $C$ is called model selection, see e.g. [9]. One particular strength of SVMs is that they can be turned in nonlinear classifiers in an elegant and effective way (see e.g. [15,9,1]).

**Fisher's discriminant**

Fisher's discriminant computes the projection **w** differently. Under the restrictive assumption that the class distributions are (identically distributed) Gaussians of equal covariance, it can be shown to be Bayes optimal. The separability of the data is measured by two quantities: How far are the projected class means apart (should be large) and how big is the variance of the data in this direction (should be small). This can be achieved by maximizing the so-called Rayleigh coefficient of between and within class variance with respect to **w** [16, 17]. The slightly stronger assumptions have been fulfilled in several of our BCI experiments e.g. in [12, 13]. When the optimization to obtain (regularized) Fisher's discriminant is formulated as a mathematical programm, cf. [18, 9, 19], it resembles the SVM:

$$\min_{\mathbf{w},b,\xi} 1/2\ ||\mathbf{w}||_2^2 + C/K\ ||\xi||_2^2 \text{ subject to}$$

$$y_k(\mathbf{w}^T\mathbf{x}_k+b)=1 - \xi_k \text{ for } k=1,\ldots,K.$$

## 2.2 Some remarks about regularization and non-robust classifiers

Linear classifiers are generally more robust than their nonlinear counterparts, since they have only limited flexibility (less free parameters to tune) and are thus less prone to overfitting. Note however that in the presence of strong noise and outliers *even* linear systems can fail. In the cartoon of Fig.2 one can clearly observe that one outlier or strong noise event can change the decision surface drastically, if the influence of single data points on learning is not limited. Although this effect can yield strongly decreased classification results for linear learning machines, it can be even more devastating for nonlinear methods. A more formal way to control one's mistrust in the available training data, is to use regularization (e.g. [20, 21, 14]). Regularization helps to limit (a) the influence of outliers or strong noise (e.g. to avoid Fig. 2 middle), (b) the complexity of the classifier (e.g. to avoid Fig. 2 right) and (c) the raggedness of the decision surface (e.g. to avoid Fig. 2 right). No matter whether linear or nonlinear methods are used, one should *always* regularize, – in particular for BCI data!

# 3 The Berlin Brain-Computer Interface

The Berlin Brain-Computer Interface is driven by the idea to shift the main burden of the learning task from the human subject to the computer under

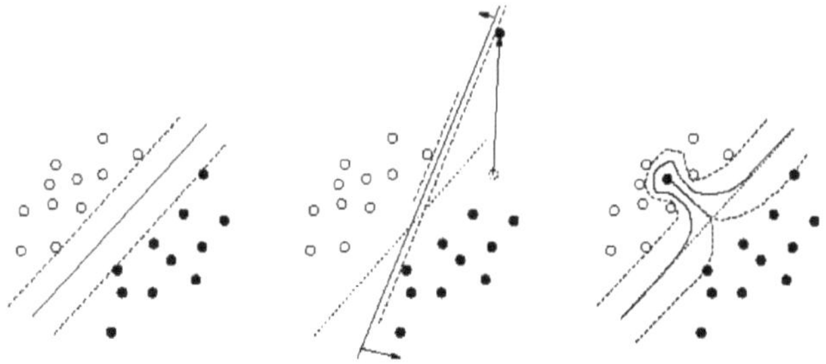

**Fig. 2.** The problem of finding a maximum margin "hyper-plane" on reliable data (left), data with an outlier (middle) and with a mislabeled pattern (right). The solid line shows the resulting decision line, whereas the dashed line marks the margin area. In the middle and on the right the original decision line is plotted with dots. Illustrated is the noise sensitivity: only one strong noise/outlier pattern can spoil the whole estimation of the decision line. From [22].

the motto 'let the machines learn'. To this end, the machine learning methods presented in the previous sections are applied to EEG data from selected BBCI paradigms: selfpaced [12, 13] and imagined [23, 24, 25] experiments.

## 3.1 Self-paced Finger Tapping Experiments

In preparation of motor tasks, a negative readiness potential precedes the actual execution. Using multi-channel EEG recordings it has been demonstrated that several brain areas contribute to this negative shift (cf. [26, 27]). In unilateral finger or hand movements the negative shift is mainly focussed on the frontal lobe in the area of the corresponding motor cortex, i.e., contralateral to the performing hand. Based on the laterality of the pre-movement potentials it is possible to discriminate multi-channel EEG recordings of upcoming left from right hand movements. Fig. 3 shows the lateralized readiness potential during a 'self-paced' experiment, as it can be revealed here by averaging over 260 trials in one subject.

In the 'self-paced' experiments, subjects were sitting in a normal chair with fingers resting in the typing position at the computer keyboard. In a deliberate order and on their own free will (but instructed to keep a pace of approximately 2 seconds), they were pressing keys with their index and little fingers.

EEG data was recorded with 27 up to 120 electrodes, arranged in the positions of the extended 10–20 system, referenced to nasion and sampled at 1000 Hz. The data were downsampled to 100 Hz for further offline analyses. Surface EMG at both forearms was recorded to determine EMG onset. In addition, horizontal and vertical electrooculograms (EOG) were recorded to check for correlated eye movements.

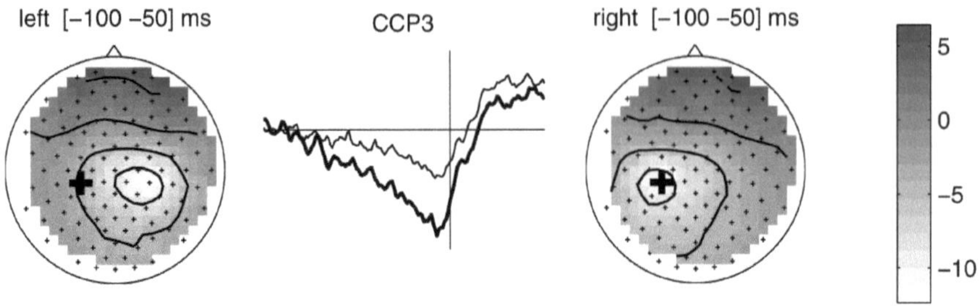


**Fig. 3.** The scalp plots show the topography of the electrical potentials prior to keypress with the left resp. right index finger. The plot in the middle depicts the event-related potential (ERP) for left (thin line) vs. right (thick line) index finger in the time interval -1000 to -500 ms relative to keypress at electrode position CCP3, which is marked by a bigger cross in the scalp plots. The contralateral negativation (lateralized readiness potential, LRP) is clearly observable. Approx. 260 trials per class have been averaged.

In [6], it has been demonstrated that when analyzing LRP data offline with the methods detailed in the previous sections, classification accuracies of more than 90% can be reached at 110 ms before the keypress, i.e. a point in time where classification on EMG is still at chance level. These findings suggest that it is possible to use a BCI in time critical applications for an early classification and a rapid response.

Table 1 shows the classification results for one subject when comparing different machine learning methods. Clearly regularization and careful model selection are mandatory which can, e.g., be seen by comparing LDA and RLDA. Of course, regularization is of more importance the higher the dimensionality of features is. The reason of the very bad performance of *k*-NN is that the underlying Euclidean metric is not appropriate for the bad signal-to-noise ratio found in EEG trials. For further details refer to [12, 13]. Note that the accuracy of 90% can be maintained in recent realtime feedback experiments [28]. Here, as no trigger information is available beforehand, the classification decision is split into one classifier that decides whether a movement is being prepared and a second classifier that decides between left and right movement to come.

### 3.2 Motor Imagery Experiments

During imagination of a movement, a lateralized attenuation of the $\mu$- and/or central $\beta$-rhythm can be observed localized in the corresponding motor and somatosensory cortex. Besides a usual spectral analysis, this effect can be visualized by plotting event-related desynchronization (ERD) curves [29] which show the temporal evolution of the band-power in a specified frequency band. A typical averaged ERD is shown in Fig. 4.

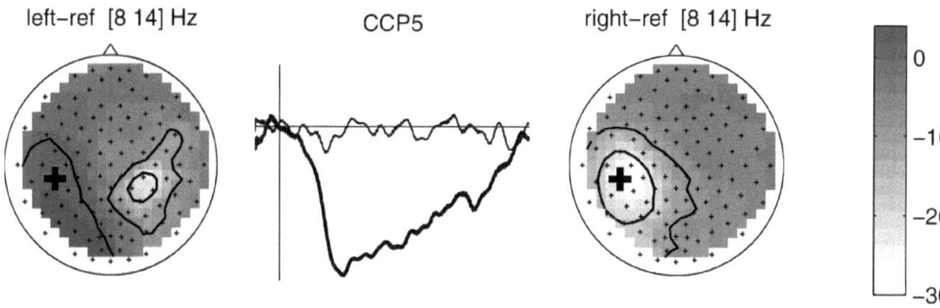


**Fig. 4.** This scalp plots show the topography of the band power in the frequency band 8–14 Hz relative to a reference period. The plot in the middle shows ERD curves (temporal evolution of band power) at channel CCP5 (mark by a bigger cross in the scalp plots) for left (thin line) and right (thick line) hand motor imagery. The contralateral attenuation of the μ-rhythm during motor imagery is clearly observable. For details on ERD, see [29].

**Table 1.** Test set error (± std) for classification at 110 ms before keystroke; ›mc‹ refers to the 56 channels over (sensori) motor cortex, ›all‹ refers to all 105 channels. The algorithms in question are Linear Discriminant Analysis (LDA), Regularized Linear Discriminant Analysis (RLDA), Linear Programming Machine (LPM), Support Vector Machine with Gaussian RBF Kernel (SVMrbf) and *k*-Nearest Neighbor (*k*-NN).

| channels | LDA | RLDA | LPM | SVMrbf | *k*-NN |
|---|---|---|---|---|---|
| All | 16.9±1.3 | 8.4±0.6 | 7.7±0.6 | 8.6±0.6 | 28.4±0.9 |
| mc | 9.3±0.6 | 6.3±0.5 | 7.4±0.7 | 6.7±0.7 | 22.0±0.9 |

We performed experiments with 6 healthy subjects performing motor imagery. The subjects were sitting comfortably in a chair with their arms in a relaxed position on an arm rest. Two different sessions of data collection were provided: In both a target "L", "R" and "F" (for left, right hand and foot movement) is presented for the duration of 3.5 seconds to the subject on a computer screen. In the first session type this is done by visualizing the letter on the middle of the screen. In the second session type the left, right or lower triangle of a moving gray rhomb is colored red. For the whole length of this period, the subjects were instructed to imagine a sensorimotor sensation/movement in left hand, right hand resp. one foot. After stimulus presentation, the screen was blank for 1.5 to 2 seconds. In this manner, 35 trials per class per session were recorded. After 25 trials, there was a short break for relaxation. Four sessions (two of each training type) were performed. EEG data was recorded with 128 electrodes together with EMG from both arms and the involved foot, and EOG as described above.

An offline machine learning analysis of the “imagined”-experiments yields again high classification rates (up to 98.9% with the feature combination algorithm PROB [24, 23]), which predicts the feasibility of this paradigm for online feedback situations (see also [25]). In fact, our recent online experiments have confirmed this prediction by showing high bitrates for several subjects. These subjects were untrained and had to play video games like ‘brain pong’, ‘basket’ (a spelling task) and ‘controlled 1-D cursor movement’ [30]. Depending on the ‘game’ scenario the best subjects could achieve information transfer rates of up to 37 Bits/min.

## 4 Conclusion

After a brief review of general linear machine learning techniques, this paper demonstrated their application in the context of real BCI-experiments. Using these techniques, it can be seen that the paradigm shift away from subject training to individualization and adaptation (‘let the machines that learn’) of the signal processing and classification algorithm to the specific brain ‘under study’ holds the key to the success of the BBCI. Being able to use (B)BCI for untrained subjects dramatically enhances and broadens the spectrum of practical applications in human-computer interfacing.

## Acknowledgments

We thank our co-authors from previous publications for letting us use the figures and joint results [1, 9, 11, 13, 22].

## References

1. Klaus-Robert Müller, Matthias Krauledat, Guido Dornhege, Gabriel Curio, and Benjamin Blankertz, “Machine learning techniques for brain-computer interfaces,” *Biomed. Technik*, vol. 49, no. 1, pp. 11–22, 2004.

2. Jonathan R. Wolpaw, Niels Birbaumer, Dennis J. McFarland, Gert Pfurtscheller, and Theresa M. Vaughan, "Brain-computer interfaces for communication and control," *Clin. Neurophysiol.*, vol. 113, pp. 767–791, 2002.
3. J.R. Wolpaw, N. Birbaumer, William J. Heetderks, D.J. McFarland, P. Hunter Peckham, G. Schalk, E. Donchin, Louis A. Quatrano, C.J. Robinson, and T.M. Vaughan, "Braincomputer interface technology: A review of the first international meeting," *IEEE Trans. Rehab. Eng.*, vol. 8, no. 2, pp. 164–173, 2000.
4. Andrea Kübler, Boris Kotchoubey, Jochen Kaiser, Jonathan Wolpaw, and Niels Birbaumer, "Brain-computer communication: Unlocking the locked in," *Psychol. Bull.*, vol. 127, no. 3, pp. 358–375, 2001.
5. Eleanor A. Curran and Maria J. Stokes, "Learning to control brain activity: A review of the production and control of EEG components for driving brain-computer interface (BCI) systems," *Brain Cogn.*, vol. 51, pp. 326–336, 2003.
6. Matthias Krauledat, Guido Dornhege, Benjamin Blankertz, Gabriel Curio, and Klaus-Robert Müller, "The berlin brain-computer interface for rapid response," *Biomed. Technik*, vol. 49, no. 1, pp. 61–62, 2004.
7. J.R. Wolpaw, D.J. McFarland, and T.M. Vaughan, "Brain-computer interface research at the Wadsworth Center," *IEEE Trans. Rehab. Eng.*, vol. 8, no. 2, pp. 222–226, 2000.
8. N. Birbaumer, N. Ghanayim, T. Hinterberger, I. Iversen, B. Kotchoubey, A. Kübler, J. Perelmouter, E. Taub, and H. Flor, "A spelling device for the paralysed," *Nature*, vol. 398, pp. 297–298, 1999.
9. K.R. Müller, S. Mika, G. Rätsch, K. Tsuda, and B. Schölkopf, "An introduction to kernel-based learning algorithms," *IEEE Transactions on Neural Networks*, vol. 12, no. 2, pp. 181–201, 2001.
10. S.S. Haykin, *Adaptive Filter Theory*, Prentice Hall, 1995.
11. Klaus-Robert Müller, Charles W. Anderson, and Gary E. Birch, "Linear and non-linear methods for brain-computer interfaces," *IEEE Trans. Neural Sys. Rehab. Eng.*, vol. 11, no. 2, pp. 165–169, 2003.
12. Benjamin Blankertz, Gabriel Curio, and Klaus-Robert Müller, "Classifying single trial EEG: Towards brain computer interfacing," in *Advances in Neural Inf. Proc. Systems (NIPS 01)*, T. G. Diettrich, S. Becker, and Z. Ghahramani, Eds., 2002, vol. 14, pp. 157–164.
13. Benjamin Blankertz, Guido Dornhege, Christin Schäfer, Roman Krepki, Jens Kohlmorgen, Klaus-Robert Müller, Volker Kunzmann, Florian Losch, and Gabriel Curio, "Boosting bit rates and error detection for the classification of fast-paced motor commands based on single-trial EEG analysis," *IEEE Trans. Neural Sys. Rehab. Eng.*, vol. 11, no. 2, pp. 127–131, 2003.
14. R.O. Duda, P.E. Hart, and D.G. Stork, *Pattern classification*, John Wiley & Sons, second edition, 2001.
15. V.N. Vapnik, *The nature of statistical learning theory*, Springer Verlag, New York, 1995.
16. R.A. Fisher, "The use of multiple measurements in taxonomic problems," *Annals of Eugenics*, vol. 7, pp. 179–188, 1936.
17. K. Fukunaga, *Introduction to Statistical Pattern Recognition*, Academic Press, San Diego, 2nd edition, 1990.

18. S. Mika, G. Rätsch, and K.-R. Müller, "A mathematical programming approach to the kernel Fisher algorithm," in *Advances in Neural Information Processing Systems*, T.K. Leen, T.G. Dietterich, and V. Tresp, Eds. 2001, vol. 13, pp. 591–597, MIT Press.
19. S. Mika, G. Rätsch, J Weston, B. Schölkopf, A. Smola, and K.-R. Müller, "Constructing descriptive and discriminative non-linear features: Rayleigh coefficients in kernel feature spaces," *IEEE Transaction on Pattern Analysis and Machine Intelligence*, vol. 25, no. 5, pp. 623–628, May 2003.
20. T. Poggio and F. Girosi, "Regularization algorithms for learning that are equivalent to multilayer networks," *Science*, vol. 247, pp. 978–982, 1990.
21. G. Orr and K.-R. Müller, Eds., *Neural Networks: Tricks of the Trade*, vol. 1524, Springer LNCS, 1998.
22. G. Rätsch, T. Onoda, and K.-R. Müller, "Soft margins for AdaBoost," *Machine Learning*, vol. 42, no. 3, pp. 287–320, Mar. 2001, also NeuroCOLT Technical Report NC-TR-1998-021.
23. Guido Dornhege, Benjamin Blankertz, Gabriel Curio, and Klaus-Robert Müller, "Increase information transfer rates in BCI by CSP extension to multi-class," in *Advances in Neural Inf. Proc. Systems (NIPS 03)*, 2004, vol. 16, in press.
24. Guido Dornhege, Benjamin Blankertz, Gabriel Curio, and Klaus-Robert Müller, "Boosting bit rates in non-invasive EEG single-trial classifications by feature combination and multi-class paradigms," *IEEE Trans. Biomed. Eng.*, vol. 51, no. 6, pp. 993–1002, 2004.
25. Matthias Krauledat, Guido Dornhege, Benjamin Blankertz, Florian Losch, Gabriel Curio, and Klaus-Robert Müller, "Improving speed and accuracy of brain-computer interfaces using readiness potential features," in *Proceedings of the 26th Annual International Conference IEEE EMBS on Biomedicine, San Francisco*, 2004, accepted.
26. R.Q. Cui, D. Huter, W. Lang, and L. Deecke, "Neuroimage of voluntary movement: topography of the Bereitschaftspotential, a 64-channel DC current source density study," *Neuroimage*, vol. 9, no. 1, pp. 124–134, 1999.
27. W. Lang, O. Zilch, C. Koska, G. Lindinger, and L. Deecke, "Negative cortical DC shifts preceding and accompanying simple and complex sequential movements," *Exp. Brain Res.*, vol. 74, no. 1, pp. 99–104, 1989.
28. Roman Krepki, Benjamin Blankertz, Gabriel Curio, and Klaus-Robert Müller, "The Berlin Brain-Computer Interface (BBCI): towards a new communication channel for online control in gaming applications," *Journal of Multimedia Tools and Applications*, 2004, invited contribution.
29. Gert Pfurtscheller and F.H. Lopes da Silva, "Event-related EEG/MEG synchronization and desynchronization: basic principles," *Clin. Neurophysiol.*, vol. 110, no. 11, pp. 1842–1857, Nov 1999.
30. Roman Krepki, Benjamin Blankertz, Gabriel Curio, and Klaus-Robert Müller, "The Berlin Brain-Computer Interface (BBCI): towards a new communication channel for online control of multimedia applications and computer games," in *9th International Conference on Distributed Multimedia Systems (DMS'03)*, 2003, pp. 237–244.

# Segmentation of Brain MR Images Using Local Multi-Resolution Histograms

Guorong Wu, Feihu Qi

Department of Computer Science and Engineering
Shanghai Jiao Tong University, Shanghai, 200030, China
grwu@cs.sjtu.edu.cn, fhqi@cs.sjtu.edu.cn

**Abstract.** Histogram based technology has been wildly used in MR image segmentation. However a single histogram suffers from the inability to encode spatial image variation. In this paper, we propose a simple yet novel classification method based on local multi-resolution histograms that utilize the difference among multiple histograms of consecutive image resolutions. We have designed such histogram based attribute vectors which have many desired properties including that they are easy to compute, translation and rotation invariant, and sufficient to encode local spatial information. Comparisons between different methods for expectation-maximization-like (EM-like) procedures and hidden markov random field method for simultaneous parameter estimation and partial volume estimation are reported. Experimental results show that our method allows for more accurate and robust classification for brain tissue than these algorithms.

## 1 Introduction

Magnetic Resonance Imaging (MRI) is currently the most reliable and accurate method for visualizing the structure of the human brain. High-resolution structural images generated with MRI display good soft tissue contrast and therefore offer the opportunity to measure the distribution of gray matter and white matter in the brain. However, the amount of data is too much for manual analysis, and this has been one of the biggest obstacles in the effective use of MRI. For this reason, automatic or semi-automatic techniques of computer-aided image analysis are necessary. Segmentation of MR images into different tissue classes, especially gray matter (GM), white matter (WM) and cerebrospinal fluid (CSF), is an important task.

The process of MR imaging procedures intensities depending on three tissue characteristics, namely: $T_1$ and $T_2$ relaxation times and proton den-

[1] This work is supported by National Science Foundation of China. No. 60271033



*G. Hommel and Sheng Huanye (eds.), Human Interaction with Machines,* 61-68.

sity (PD). The effect of these parameters on the images can be varied by adjusting parameters such as time to echo (TE) and time to repeat (TR) of the pulse sequence. By using different parameters or numbers of echoes in the pulse sequence, a multitude of nearly registered images with different characteristics of the same object can be achieved. If only a single MR image of the object is available, such an image is referred to as single-channel and when a number of MR images of the same object at the same section are obtained, they are referred as multi-channel images. Some segmentation techniques have relied on multi-spectral characteristics of the MR image, while in this paper we present an automated segmentation method for single-channel MR images which have been stripped the skull.

Because of the finite resolution of the imaging devices, a single voxel may contain several tissue types. This is known as partial volume effect (PVE) [1]. Due to PVE, the classification of a voxel reflecting the dominant tissue type (WM, GM, or CSF), does not reveal the all possible information about the tissue content of that voxel. This can be problematic in small structures or highly convoluted areas of the brain. Partial volume effect and PV estimation have been addressed in various ways in MR imaging literature. For example, Pham and Prince [2] have proposed a fuzzy C-means algorithm with the relationship between the fuzzy C-means objective function and statistical models of PVE in simplified case; Wang *et al* [3] proposed to use a Bayesian classifier with a variable number of tissue classed, including classes of mixed tissue types. The most commonly used model of PVE is the mixel model proposed by Choi *et al* [4]. This approach assumes that each intensity value in the image is a realization of a weighed sum of random variables, each of which characterizes a pure tissue type. The method involves maximum-likelihood estimation of the partial volume coefficients for each voxel that model partial volume fractions of pure tissue type. In general, there are three approaches to parameter estimation problem: histogram analysis, simultaneous parameter, and partial volume estimation by expectation maximization (EM)-like algorithms, and estimation based on a hard segmentation of the image.

Histogram based methods are widely used among these three approaches. Grabowski *et al* [5] proposed to fit Gaussian distributions to global and local voxel intensity histograms. They assume gaussian distribution for each of the basic tissue types and two mixture gaussian distributions for partial gray/white and CSF/gray, respectively. However, histogram based methods have their drawbacks. Histogram analysis requires a mixture probability density to be fit to an image histogram by parameter optimization. This involves finding the minimizer of a multimodal objective function and therefore reliability of histogram analysis for parameter estimation depends heavily on the optimization algorithm used for the fitting task. If a standard nonlinear optimization algorithm aimed at local minimization (e.g. Levenberg-Marquee used commonly for curve fitting) is used, the initialization for the algorithm has to be chosen carefully to avoid convergence to a poor local minimum. These considerations call for the use of advanced global optimization algorithms, but the problem with global optimization methods is that they are usually far more time consuming than local optimization methods.

Hadjidemetriou [6] proposed to use multi-resolution histograms for texture image recognition. They compute the histogram of multiple resolutions of an image to form a multi-resolution histogram. The multi-resolution histogram shares many desirable properties with the plain histogram including that they are both fast to compute, space efficient, invariant to rigid motions, and robust to noise. In addition, the multi-resolution histogram directly encodes spatial information. In this paper, we extend this feature to local multi-resolution histogram of MR brain images. We design a new attribute vector of each voxel within the local histograms from different resolution and utilize these attribute vectors instead of intensity to estimate the partial volume coefficients. The paper is organized as follows: section 2 describes the mathematical background of the multi-resolution histogram and their extension in MR image segmentation. Section 3 gives experimental results with comparison of other EM-like methods. Section 4 we will discuss the results and limitations of our segmentation scheme.

# 2 Methods

## 2.1 Multi-resolution histogram and generalized fisher information measures

The multi-resolution decomposition of an image is computed with gaussian filtering. The image at each resolution gives a different histogram. The multi-resolution histogram, $\boldsymbol{H}$, is the set of intensity histogram of an image at multiple image resolutions. Translations, rotations, and reflections preserve the multi-resolution histogram. In general, however, the transformations of an image affect its multi-resolution histogram. This effect is addressed using as analytical tools the image functional called generalized Fisher information measures. These functional relate histogram density values with spatial image variation.

The Tsallis generalized entropies of an image $\boldsymbol{L}$ depend on a continuous parameters $\boldsymbol{q}$ and are given by

$$S_q = \int_D \frac{L(x) - L^q(x)}{q-1} d^2x \tag{1}$$

where image ***L*** has unit $L_1$ norm and ***L(x)*** is the intensity value at pixel x. Note, we consider 2D situation for discuss convenience while true implemented in 3D. In the limit $q \to 1$ the Tsallis generalized entropies reduce to the Shannon entropy. In equation (1), the intensities at all points ***x***, denoted by ***L(x)*** can be substituted directly by their values: $v_0, v_1, \ldots v_{m-1}$, where ***m*** is the total number of gray levels. The union of all the regions in the domain with identical intensity, $v_j$, gives the value of histogram density $j$, $h_j$. That is, equation (1) becomes

$$S_q = \sum_{j=0}^{m-1} \left( \frac{v_j - v_j^q}{q-1} \right) h_j \tag{2}$$

That is, the Tsallis generalized entropies can be expressed as a linear function of the histogram.

To decrease image resolution, we use a gaussian filter ***G***(***l***):

$$G(l) = \frac{1}{2\pi l \sigma^2} \exp\left( -\frac{x^2 + y^2}{2l\sigma^2} \right) \tag{3}$$

where $\sigma$ is the standard deviation of the filter, and ***l*** denotes the image resolution. A filtered image, ***L*G***(***l***), has histogram ***h***(***L*G***(***l***)) and entropy vector ***S***(***L*G***(***l***)). The rate at which the histogram changes with image resolution can be related to the rate at which image entropies change with image resolution. This relationship is obtained by differentiating equation (2) with respect to ***l*** to obtain

$$\frac{dS_q(L*G(l))}{dl} = \sum_{j=0}^{m-1} \left( \frac{v_j - v_j^q}{q-1} \right) \frac{dh_j(L*G(l))}{dl} \tag{4}$$

the rate at which the Tsallis generalized entropies change with image resolution ***l*** on the left hand side of equation (4) are related to closed form functional of the image. This functional are the generalized Fisher information measures

$$J_q(L) = \frac{\sigma^2}{2} \frac{dS_q(L*G(l))}{dl} \tag{5}$$

The substitution of equation (4) into the right-hand side of equation (5) gives

$$J_q(L) = \frac{\sigma^2}{2} \sum_{j=0}^{m-1} \left( \frac{v_j - v_j^q}{q-1} \right) \frac{dh_j(L*G(l))}{dl} \tag{6}$$

This equation reveals that $J_q$ is linearly proportional to the rate at which the histogram densities change with image resolution. The proportionality factors in equation (6), for $q$>1 weigh heavier the rate of change of the histogram densities that have large intensity values and vice-versa. The proportionality factors of the Fisher information, $J_1$, weigh approximately equally all histogram densities. The rate of change the histogram with image resolution can be transformed into the $m$*1 vector $J=[J_{q0}\ J_{q1}\ J_{q2}\ldots J_{qm1}]^T$. The component of vector $J$ that will be used to a large extent is the Fisher information, $J_1$.

## 2.2 Classification using local multi-resolution histogram

Multi-resolution histogram has many advantages and has been successfully used for texture image classification. We extent these features to 3D MR brain image segmentation. Our contributions include: **1.** In texture image classification, they deal with 2D images, while we deal with 3D MR brain images. The latter is more complex than the former; **2.** The multi-resolution histogram of the whole image was computed in 2D texture image classification, that is, one image has an unique value of Fisher information measure $J_q$, while in our case we extract local multi-resolution histogram of each voxel and get the $J_q$ value of that voxel within the local histogram; **3.** We extract the local histogram with adaptive neighborhood size. That means the size of neighborhood is variable at different regions; **4.** We design a new classification algorithm which uses the new attribute vector instead of intensity.

One of our extensions is to extract the local multi-resolution histograms of each voxel. In practice, we convolve the image with three Gaussian filters and obtain three new image of different resolution called high-resolution image, middle-resolution image, and low-resolution image. The standard variances of Gaussian filters are 0.5, 1.0, and 2.0 respectively. At each voxel, the value of Fisher information measure between high-resolution image and middle resolution image $J_1^{\text{high-mid}}$ and the value of Fisher information measure between middle resolution image and low-resolution image $J_1^{\text{mid-low}}$ have been computed according to the local multi-resolution histograms.

In addition, we compute the Fisher information measure with different neighborhood size at each voxel. Intuitively, we should use large neighborhood size at the region where the image information is poor while use small neighborhood size at the region where the image information is rich. Our method is to decide the neighborhood size according to image intensity entropy. The larger value of entropy, the larger neighborhood size the voxel has. More heuristic method will be of our future work.

Our classification method in fact is a machine learning procedure. At each voxel, we define a attribute vector: $\boldsymbol{AV}$=($\boldsymbol{v}$, $\boldsymbol{J}_1^{\text{high-mid}}$, $\boldsymbol{J}_1^{\text{high-mid}}$), where $\boldsymbol{v}$ is intensity at that voxel. It is reasonable to assume that voxels lying on the boundaries between tissue types in hard-segmented image are likely to contain partial volume effect. For this reason, voxels with at least one of their 6 neighbors belonging to a different class than the voxel itself are considered to be PVE regions. First, we choose some training images and pick out the PVE region according to some hard-segmented algorithm (K-means clustering method). Then we compute the attribute vectors defined above of those regions. Next, we train the SVM with these attribute vectors. In the classification stage, we still first obtain the PVE regions according to hard-segmented result and compute the attribute vectors of each voxel. SVM is used to predict the final labeling of each voxel. Details of our classification algorithm are listed as follows and the graphical pipeline demonstration of our method is displayed in figure. 1.

**Training Stage:**

Step1: use Tree Structured K-Means algorithm (TSK) to get the hard-segmented image;

Step2: pick out the regions involving PVE and compute the attribute vectors of each voxel in those regions;

Step3: train the SVM with those attributes vectors.

**Classification Stage:**

Step1: perform TSK algorithm;

Step2: compute attribute vectors at the located PVE regions;

Step3: use SVM to predict the label according to the attribute vector

Step4: Smooth the label image. If the convergence condition is not satisfied, go to step 1

Notably, the termination condition we used is the number of flips of voxel label in PVE regions. In experiment, we select 0.1% of total number of voxels in PVE regions as the threshold.

## 3 Results

We have applied our new technique to real MR datasets. Each volume consists of 256×256×124 voxels and each voxel is 0.9375×0.9375×1.5 $mm^3$. We compare our proposed method with three well known segmentation algorithm: AMAP (Adaptive MAP) [7], BMAP (Biased MAP) [8], HMM [9] (Hidden Markov Model). All of these three methods build on statistical model and estimate the parameters using MAP (Maximum *a prior*) or ML (Maximum Likelihood) principle. The results are show in figure. 2. From left to right are intensity image, segmentation result of AMAP (Fig. 2b(1)), BMAP (Fig. 2b(2)), HMM (Fig2. b(3)), and our results (Fig. 2b(4)). The red ellipses represent the most salient PVE regions. It is obvious for us to find out lenticular and caudate around the ventricular. AMAP, BMAP, HMM algorithm manipulate on voxel intensities, and at PVE regions, white matter has dominated the gray matter. While our method not only use voxel intensity but also the local multi-resolution histograms. Our result is much better than previous three algorithms. In addition, our segmentation result of caudate and lenticular is validated by medical experts.

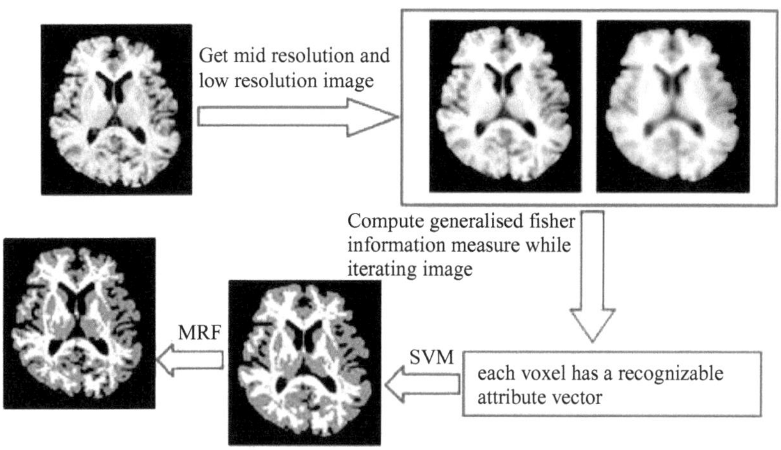


**Fig. 1.** Graphical pipeline of our classification method.

## 4 Conclusion

In this paper, we proposed a new classification method of MR brain images using local multi-resolution histograms. The local multi-resolution histograms capture more spatial information than single voxel intensities and it retains simplicity, efficiency, and robustness of the plain histograms. The experimental results show that our method gain more accurate result in PVE regions than parameter estimation like methods.

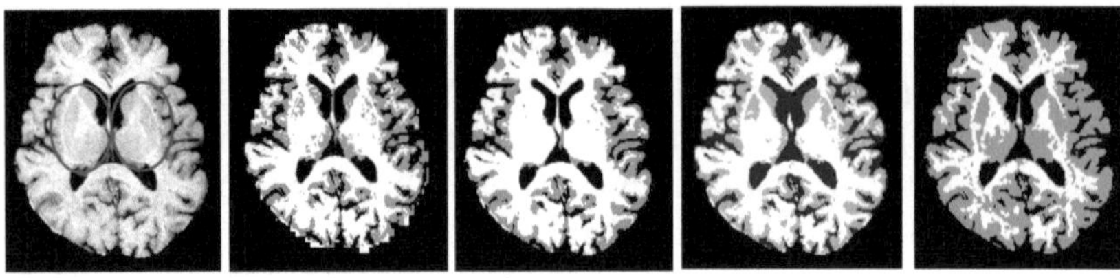


**Fig. 2.** Results of classification method. Images from left to right are intensity image, classification result of AMAP, BMAP, HMM, and our results. The red ellipses represent the most salient PVE regions.

## References

[1] J. Tohka, A. Zijdenbos, and A. Evans, "Fast and robust parameter estimation for statistical partial volume models in brain MRI", NeuroImage, vol. 23, pp. 84-97, 2004.

[2] Pham, D.L., Prince, J., "Adaptive fuzzy segmentation of magnetic resonance image.", IEEE Transaction on Medical Imaging 18(9), pp. 737-752, 1999.

[3] Wang, Y., Adah, T., Xuan, J., "Magnetic Resonance image analysis by information theoretic criteria and stochastic site model", IEEE Transaction on Information Technology Biomedicine, 5(2), pp. 150-158, 2001.

[4] Choi, H.S., Haynor, D.R., Kim, Y., "Partial volume tissue classification of multichannel magnetic resonance images – a mixel model", IEEE Transaction Medical Imaging, Vol(10), No. 3, pp. 395-407, 1991.

[5] T.J. Grabowski, R.J. Frand, N.R. Szumski, C.K. Brown, and H. Damasio, "Validation of partial tissue segmentation of single-channel magnetic resonance images of the brain", NeuroImage, vol. 12, pp. 640-656, 2000.

[6] E. Hadjidemetriou, M.D. Grossberg, S.K. Nayar, "Multiresolution histograms and their use for recognition", IEEE Transactions on Pattern Analysis and Machine Intelligence, vol. 26, No. 7, 2004.

[7] J.C. Rajapakse, J.N. Giedd, J.L. Rapoport, "Statistical approach to segmentation of single-channel cerebral MR Images", IEEE Transaction on Medical Imaging vol. 16, No. 2, pp. 176-186, 1997.

[8] J.C. Rajapakse, F. Druggel, "Segmentation of MR images with intensity inhomogeneties", Image Vision Computing, vol. 16, pp. 165-180, 1998.

[9] Y. Zhang, M. Brady, and S. Smith, "Segmentation of brain MR images through a hidden markov random field model and the expectation-maximization algorithm", IEEE Transaction on Medical Imaging, vol. 20, No. 1, pp. 45-57, 2001.

# EMG-Driven Human Model for Orthosis Control

Christian Fleischer, Günter Hommel

Institute for Computer Engineering and Microelectronics
Berlin University of Technology, Germany
{fleischer, hommel}@cs.tu-berlin.de

**Abstract.** In this paper we present a simplified body model of the human lower extremities used for the computation of the *intended motion* of a subject wearing an exoskeleton orthosis. The *intended motion* is calculated by analyzing EMG signals emitted by selected muscles. With the calculated *intended motion* a leg orthosis is controlled in real-time performing the desired motion.

To allow motions with different velocities and accelerations, the body model contains physical properties of the body parts and is animated with data recorded from the pose sensors as a basis for the prediction. Computing the *intended* motion is achieved by converting calibrated EMG signals to joint torques and forces which are also part of the model. The extrapolation is performed for a short period of time, calculating the joint coordinates for the actuator control loop.

The algorithm was examined with the experiment of flexing and extending the knee while raising and lowering the thigh. The discussion compares the motion performed by the leg orthosis and the desired motion.

The algorithm of the model and the preliminary experimental results are both presented.

## Introduction

Beside the standard application of EMG signals to analyse disabilities or to track progress in rehabilitation, more focus has been put on controling robot arms and exoskeletons with EMG signals (Lee (1984), Fukuda (1999), Mirota (2001)) in recent years. In Lloyd (2003) a promising but very complex musculoskeletal model is presented that takes into account 13 muscles crossing the knee to estimate the resulting knee torque.

The advantage of EMG signals is that they form an intuitive interface and they can be used with every patient who is not paralyzed. Even if the muscles are not strong enough or the limbs hindered while performing a motion, signals of the *intended motion* (desired motion that cannot be performed) can still be recorded. In our environment the orthosis (see Figure 1) that is attached to the leg restricts the motion in the knee if the ac-



*G. Hommel and Sheng Huanye (eds.), Human Interaction with Machines,* 69-76.

tuator is not powered, so the intention has to be detected without the possibility of detecting any motion.

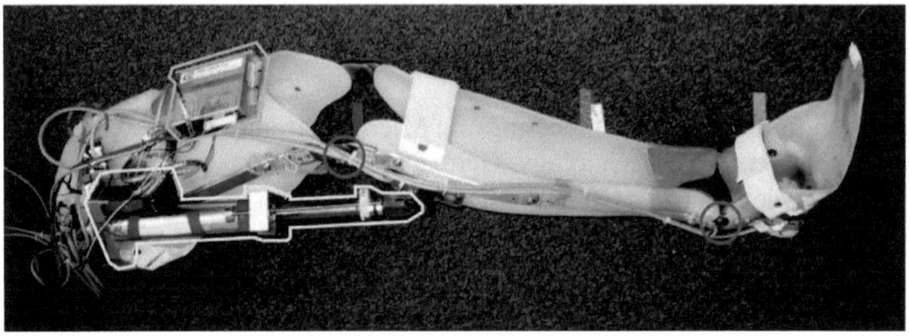

**Fig. 1.** The orthosis for the right leg. Hall sensors are marked with solid circles (red), accelerometers with dashed circles (red), the actuator (yellow) and servo amplifier (green).

## System Description

As can be seen in Figure 2, the system is divided into two parts, the kernel block with real-time data acquisition together with the PID-controller for the actuator and the motion analysis block with the biomechanical model in the user space.

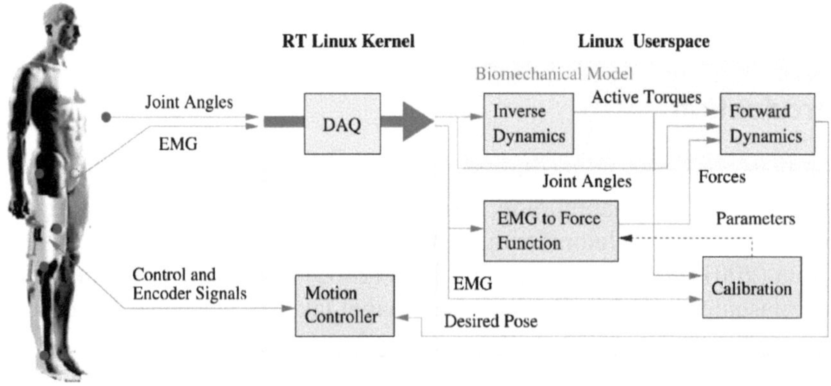


**Fig. 2.** System overview. The data are recorded from the pose and EMG sensors attached to the orthosis and passed to the biomechanical model to predict the intended motion. The result is fed into the motion controller to move the orthosis to the desired pose.

## Data Acquisition System

The measurement system used for the algorithm consists of two groups of sensors: The EMG sensors to read the muscle activity and the pose sensors to get the current state of the subject.

The EMG sensors are placed on top of two muscles responsible for flexing and extending the knee: the M. semitendinosus and M. rectus femoris. Many other muscles cooperate during this motion but we have chosen the ones with the largest contributions to the resulting torque in the knee in our setup (see Platzer 2003). The signals are sampled from DelSys 2.3 Differential Signal Conditioning Electrodes.

Ankle and knee angles are measured on both legs with Philips KMZ41 hall sensors and the thigh and trunk angles with accelerometers ADXL210 from AnalogDevices Inc. (as described in Willemsen (1991) and Fleischer (2004)), all only in the sagittal plane. All sensors are sampled with 1 KHz.

## Signal Flow

As mentioned in the introduction, the intended motion of the subject should be analysed to let the *human* control the *orthosis*. To be able to compute the *desired pose*, the *current pose* of the subject is read from the *pose sensors* attached to the limbs and fed into the *biomechanical model* together with the EMG signals from the appropriate muscles. The *biomechanical model* then calculates the *desired pose* for the next timestep and passes it to the *motion controller* that is responsible for controlling the actuator towards the desired pose.

To be able to use the *EMG-to-force function*, parameters of the function have to be calibrated. This is performed in the block *calibration*: The *biomechanical model* calculates through inverse dynamics the *active forces* (forces that must have been active in the joint crossed by the modeled muscles that resulted in the latest motion). Those forces (for the knee extensor the knee torque is used) are fed into the block *calibration* together with the corresponding EMG values to optimize the parameters of the EMG-to-force function.

## Human Body Model

The human body model consists of two legs with feet, shanks, thighs and the torso. All limbs and the torso are modeled as rigid bodies (rectangular parallelepipeds) connected with swivel joints. Body masses are calculated as fixed fractions of the total body weight ($m_{total}$=88 kg) of the subject (the figures can be found in Winter (1990)). Body dimensions are taken from our subject. Two muscles $M_f$ and $M_e$ have been added producing the corresponding force $F_{Mf}$ and torque $T_{Me}$ to allow flexion and extension of the knee (due to the anatomy of the knee extensor in the regarded range of motion it is better to use the torque here). The points of origin and insertion of $M_f$ are fixed and have been chosen by hand in analogy to human anatomy. Furthermore, the model takes into account ground reaction forces at both feet and gravity. The generalized velocities **u** and accelerations $\dot{\mathbf{u}}$ are calculated as derivations of the generalized coordinates **q** (angles recorded from the pose-sensors).

The dynamic equations of the body model have been generated with the symbolic manipulation tool AUTOLEV. Details of the model and the calibration algorithm can be found in Fleischer (2004).

## Calibration

The calibration algorithm takes pairs of post-processed EMG values and muscles forces calculated by the inverse dynamics from the same point in time and stores them in a table indexed by the EMG value: the activation level of the muscle. Older values might be overwritten. When the calibration is performed, all pairs stored in the table are taken as points on the EMG-to-Force function

$$F(x) = a_0 \cdot (1 - e^{-a_1 \cdot x}) + a_2 .$$

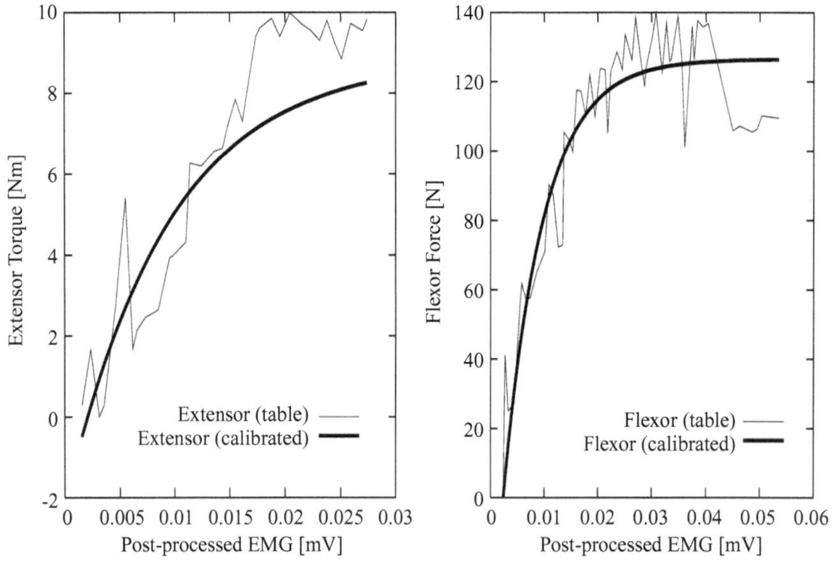


**Fig. 3.** This diagram shows the results from the calibration of the EMG-signals for the knee flexor and extensor: The Functions $F_e(x)$ and $F_f(x)$.

The Nelder-Mead Simplex algorithm is used to optimize the parameters $a_0$, $a_1$ and $a_2$ by minimizing the least square error of the force from the function F(x) and the force value stored in the table.

In Figure 3 the contents of the table is plotted as a function of the EMG signal together with the calibrated functions $F_e(x)$ and $F_f(x)$ for the knee extensor and knee flexor.

## Motion Prediction

During initialization of the algorithm the body model is synchronized with the current state **S**=(**q**, **u**) of the subject. For the knee joint the dynamic equations of the model are solved for acceleration $\dot{u}_{knee}$ and computed by applying the EMG signals to muscles $M_f$ and $T_{Me}$. Both $M_f$ and $T_{Me}$ are greater or equal zero, co-contraction of the muscles is allowed here. Only during calibration this is impossible. After double-integrating the acceleration of the knee joint for one timestep $\Delta t$=10ms into the future we get the desired angle from $\mathbf{S}_{t+\Delta t}$. Obviously this could be done for more joints (e.g. hip, ankle), but in our experiments currently only the knee joint is powered, thus we only need to compute $q_{knee}$.

## Experiments

The experiment described here was performed in an upright standing position with the left foot on the ground. The right thigh and shank have been raised and lowered in a random pattern in sagittal plane as shown in Figure 4. When interpreting the diagram it is important to take into consideration the hip angle, which is also shown there. In the first case the actuator was not attached to the orthosis so that unhindered movement was possible. As can be seen, the post-processed EMG signals of the knee flexor and extensor lead to a knee angle prediction similar in shape to the performed motion. The maximum error is 15.4 degrees, the average of the absolute error is 4.9 degrees and the standard deviation of the error is 5.9 degrees. The relative error is not meaningful since the amount is important independently of the angle where it appears. The shifting in time is mostly a result of the low-pass filtering of the EMG values and a simple friction function that is used to simulate the effects of tendons and joint end stop.

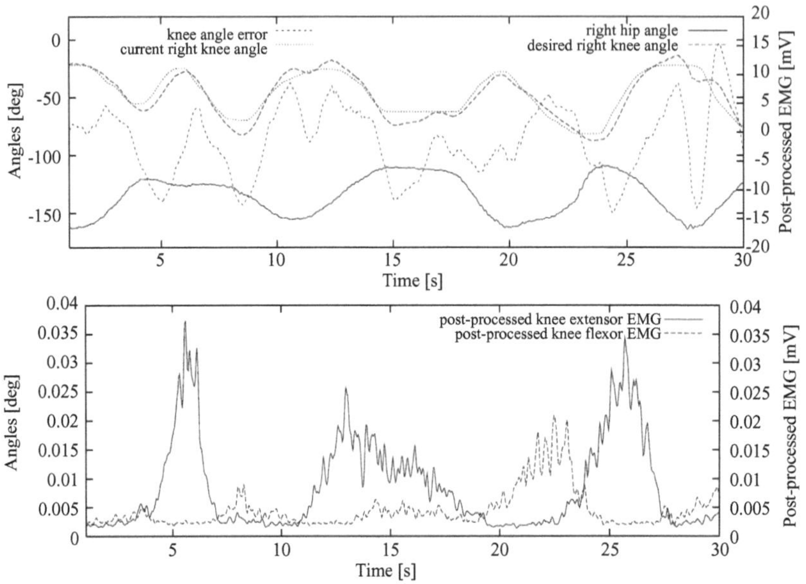


**Fig. 4.** This diagram shows the angles of the right hip and right knee with the actuator not attached. Zero degrees means upright standing position, positive angles stand for hip flexion and knee extension. Also shown is the error between the current and predicted knee angle and the post-processed EMG-values feeding the prediction for the knee muscles.

In the second experimental setup the EMG signals have been calibrated during unhindered motion at the beginning. After that, the actuator has been attached to the orthosis and powered by the motion controller. Results from this setup can be seen in Figure 5. Due to safety reasons the output current of the servo amplifier was limited resulting in limited accelerations of the actuator. Also some peak angles could not be reached due to lack of resulting knee torque. Unfortunately, due to the delayed response of the system to the desired angles, a subjective feeling of stiffness was perceived making it a little bit harder to perform the motion than without the actuator. Due to the difference in the performed motions (slight differences in angle configurations or velocities are sufficient), the EMG signals cannot be compared easily to show this effect. Additional force sensors are necessary to detect the forces of the human leg acting on the orthosis (in the case of no external contact of the orthosis).

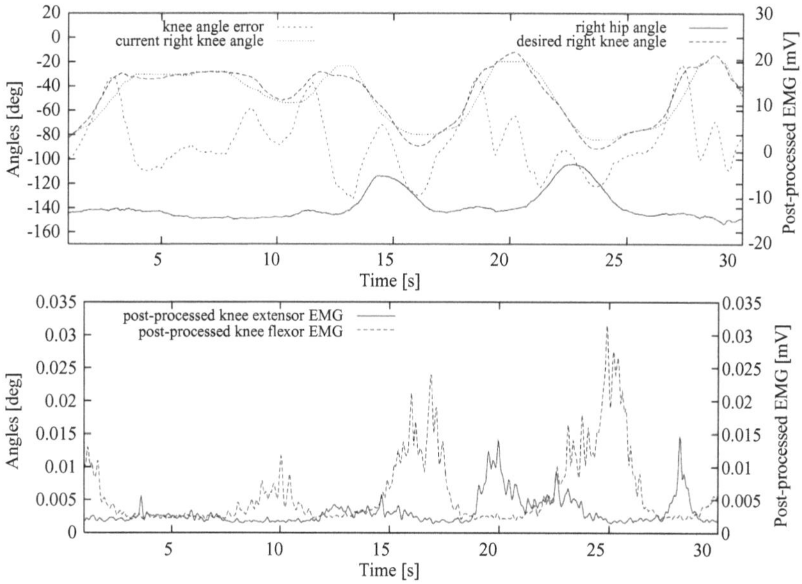


**Fig. 5.** This figure shows the motion of the right leg with the actuator attached and powered by the motion controller. Due to safety reasons the maximum current to the motor was limited resulting in a slower response to the desired angle.

## Discussion

In this paper an approach has been presented and experimentally examined that allows the estimation of the intended motion of a subject wearing a leg orthosis by evaluating EMG signals from certain muscles.

As shown in the previous section, the motion prediction algorithm is working well in predicting the desired motion, although the raw and post-processed EMG signals are quite unsteady and unreliable by nature. As was explained before, the delay between raw EMG signals and the resulting prediction for the knee torque has a strong effect on the performance of the system. The next steps in research will be to shorten this delay to allow the orthosis to be more supportive and to incorporate more muscles crossing the knee to make the model robust against contact from the environment while walking or climbing stairs.

Hopefully in the near future this research will lead to an intuitive human-to-robot interface to control powered orthoses.

## References

Fleischer C, Kondak K, Reinicke C, Hommel G (2004) Online Calibration of the EMG Force Relationship, Proceedings of the IEEE/RSJ Int. Conf. on Intelligent Robots and Systems.

Fukuda O, Tsuji T, Shigeyoshi H, Kaneko M (1999) An EMG Controlled Human Supporting Robot Using Neural Network, Proceedings of the IEEE/RSJ Int. Conf. on Intelligent Robots and Systems}, pp. 1586-1591.

Lee S, Sridis GN (1984) The Control of a Prosthetic Arm by EMG Pattern Recognition, IEEE Transactions on Automatic Control, vol. AC-29, pp. 290-302.

Lloyd DG, Besier TF (2003) An EMG-driven musculoskeletal model to estimate muscle forces and knee joint moments in vivo, Journal of Biomechanics, vol. 36, pp. 765-776.

Morita S, Kondo T, Ito K (2001) Estimation of Forearm Movement from EMG Signal and Application to Prosthetic Hand Control, Proceedings of the IEEE Int. Conf. on Robotics & Automation, pp. 3692-3697.

Platzer W (2003) Taschenatlas der Anatomie. Thieme.

Willemsen ATM, Frigo Carlo, Boom HBK (1991) Lower Extremity Angle Measurement with Accelerometers - Error and Sensitivity Analysis. IEEE Transactions on Biomedical Engineering, vol. 38, number 12.

Winter DA (1990) Biomechanics and Motor Control of Human Movement, John Wiley & Sons, Inc.

# Virtual Information Center for Human Interaction

Fang Li, Hui Shao, Huanye Sheng

Dept. of Computer Science & Engineering
Shanghai Jiao Tong University, Shanghai 200030 China
[fli, thomasshao, hysheng]@sjtu.edu.cn

**Abstract**: The paper describes the background, aim and realization of virtual information center on the domain of computational linguistics. The Internet provides huge amount of heterogeneous data sources. Human can access all information on the Internet by mouse-clicks. Our virtual information center has been established in order to provide an integrated view on computational linguistics. It consists of three kinds of data: knowledge and information, products and resource, research groups and experts. The main features for human interaction are virtual information management, distributed access, entity-based information retrieval and automatic hyperlink.
**Keywords:** Language resource, Web-based database, automatic hyperlink

## 1 Introduction

The Internet has become a rich information resource. Every day, huge amount of people searches the Internet for information of interest such as all kinds of news, goods, papers and so on. However, users cannot acquire exact information correctly and quickly using existing search engines, because of keywords ambiguity. It results low precision and low recall of information retrieval.

Computational Linguistic is a discipline between linguistics and computer science, which is concerned with the computational aspects of the human language faculty. It belongs to the cognitive sciences and overlaps with the field of ***artificial intelligence (AI)***, a branch of ***computer science*** aiming at computational models of human cognition [Uszkoreit 2002]. Due to cross-discipline, different methods, different research institutes, it is hard to share information and resources during the research, which may induce waste of money and human labor. In order to solve the problem, we need a system, which can locate all resources related to computational linguistics,



*G. Hommel and Sheng Huanye (eds.), Human Interaction with Machines,* 77-83.

categorize them, help different users to find his needed information on the domain and provide an integrated view on computational linguistics. Those resources including experts, projects, products and so on, still belong to themselves, maintained by them. This is the essence of virtual information center.

In the following, we describe the research background, then the realization of virtual information center, and finally the maintenance and our future work for the virtual information center.

## 2 Research background & Overview

As the language resources, research institutes increase with the emergence of many related technologies, It becomes more and more difficult for searching existing resources, tools and so on. The Open Language Archives Community [Bird and Simons 2003] has defined the standard architecture for different data and their related data resource in order to exchange all kinds of language resources. The main element is XML-based format. It combines the standard of resource description model DCMS (the Dublin Core Metadata Set) and common format for electronic document archive OAI (the Open Archives Initiative)[Simons and Bird 2003]. Based on such model, the virtual information center for language technology (LT-world: www.lt-world.org) [Capstick et al 2002] is established to provide all kinds of information in the domain of language technology. Using the same idea, we realize a virtual information center, which collect those projects, tools, and experts on computational linguistics in China, as a supplement for LT-world. [Li et al 2003]

There are four kinds of information in our virtual information center:

- Knowledge, terms and information: many branches of computational linguistics.
- Products, literatures and resources: already existed or developed research products, literatures and resources in China.
- Research groups, projects and experts: many groups, projects and experts in China on this area.
- Experimental system demo and evaluations: provide a window to online demo and evaluate experiment systems. (under building)

The four kinds of information are correlated with each other, for example, one expert may conduct several projects in different branches of computational linguistics. All experts, projects and terms, which describe their research areas, are connected by hyperlinks. Through terms, experts and projects can also be reached by hyperlinks. The whole architecture is in the fig.1.

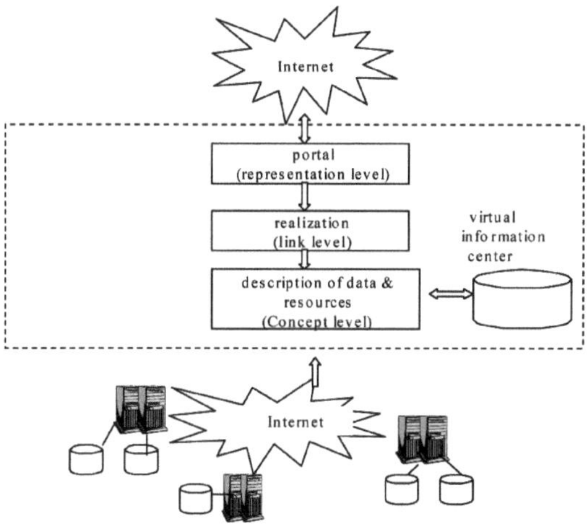


**Fig. 1.** The architecture of virtual information.

There are three levels in the virtual information center: The concept level defines the content of virtual information center with an integrated, systematical way. It is based on their independencies and relations of individual entities which can be experts, resources or projects; the representation level is in fact a portal to provide a user an interface to search all kinds of information on Computational linguistics; the link level is the realization of connecting user's queries with the database implemented with PHP, MySQL and Apache (Web Server). The realization is described in the following section.

## 3 Realization of the Virtual Information Center

The virtual information center with four kinds of information is shown in fig. 2. Human beings who search the Internet with the help of many search engines at the beginning time collect the information of database. Now we have more than one hundred of experts, projects information of China, more than three hundreds of literatures, products and resources in the database. The virtual information center can be accessed by a link through our department web site: http://www.cs.sjtu.edu.cn. In future we will collect more and update data incrementally.

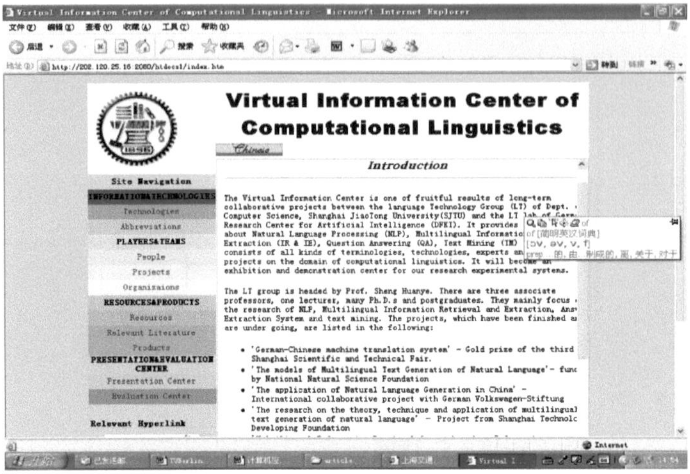


**Fig. 2.** The entry of virtual information center.

### 3.1 Concept Level

First we define concepts in the domain of computational linguistics. Concepts are those technical terms, which reflect some research branches. There are ten subjects; each subject has also many branches. For example, subject: research on language resource, which includes text corpora, lexicons, tagged corpus, grammar, synonym, multilingual corpora and so on.

The database consists of many entities: experts, research teams, projects, products, and literatures. Each entity has its own attributes, such as, name, title, affiliation, address, and homepage for experts. There are also many relationships between those entities. Instead of creating a relationship table in traditional database design, hyperlinks are used to establish relationships among those entities. For example, a relationship between experts and projects: some experts may lead or join some projects; on the other hand, a leader of a project is also an expert. When you view the information of an expert, his research projects can also be accessed through hyperlink. While searching information of projects, related experts information may also be reached by mouse-click.

All entities are connected not only together, but also with concepts (technical terms) on the domain of computational linguistics. All entities (experts, teams, projects and so on) are belong to some branches represented by terms, that means, according to the terms, you can find experts, projects, products and literatures of this research branches represented by terms. Figure 3 is an example. We use a string-matching algorithm to create hyperlinks to establish all relationships between entities.

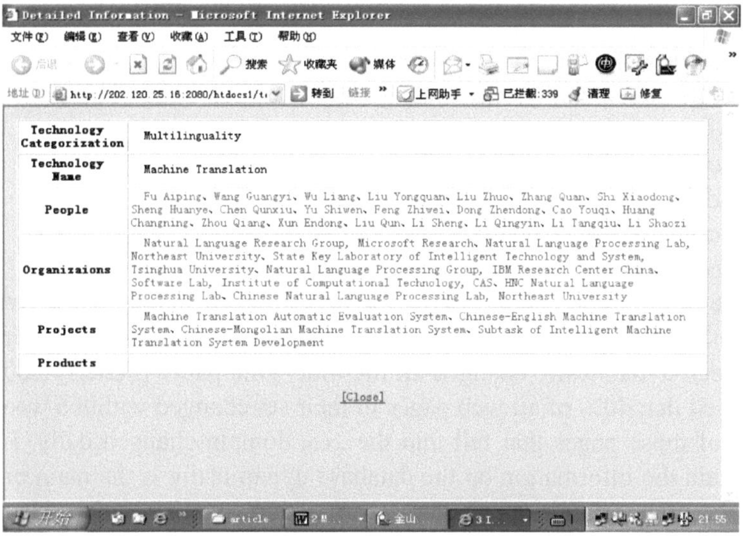


**Fig. 3.** Hyperlink Generation based on Terms.

## 3.2 Representation level

The aim of representation level is to let user access the virtual information by the Internet conveniently. The design of user interface is the web page design. It consists of static web pages and dynamic web pages. The entry of the site is a static web page, which shows some basic functions of the virtual information and provides the interface with human interaction. There are two kinds of dynamic web pages:

- The result of information retrieval after a user inputs keywords

  The following four steps are the process to get a result of information retrieval:

  1. Acquire user input keywords through the web page.
  2. Create a connection to the database using mysql_connect() command.
  3. Make database queries using mysql_db_query().
  4. Show results of database queries.
  5. Pages generated by hyperlinks.

- Pages generated by hyperlink

  While the information of database is viewed, some values are matched by hyperlinks, which are realized by a special string marching algorithm.

### 3.3 Maintenance of the Virtual Information Center

In order to facility the maintenance of database, we realized a small database management system using ADO of Visual C++. The functions are listed in the following:

1. Access Database through ODBC.
2. Update contents of database.
3. Query database through one or more tables.
4. Import batch data from outside.

The information is collected at the beginning by human beings with the help of many search engines on the Internet. However, data on the Internet has been dynamically changed all the time. One paper [Fetterly 2003] has reported that 40% of all web pages in their set changed within a week, and 23% of those pages that fell into the .com domain changed daily. How to maintain the information on the database dynamically is the main problem to face after establishment of the virtual information center. Man-made maintenance is far from keeping data updated. The future efforts will be focused on realization of dynamic and automatic maintenance. We have already realized an experimental system for personal information extraction from their homepages in order to update the expert's information of database automatically. The validation of result and final update still need human interaction.

## Conclusion

In this paper, we describe our virtual information center for computational linguistics. The virtual information center can provide an integrated view on specific domain and make the information conceptualized and centralized. Distributed access and management, integrated view are the properties of virtual information center. Much more work need to be done in the future, such as dynamical maintenance and update, active searching new information in computational linguistics.

## Acknowledgment

We are grateful to all students who have helped to create the virtual information center. Special thanks are given to Mr. Zhu Xiaoming, Zhang Wenjie, Yuan Shuangqing, and Lu Cuiming.

## References


1. Dennis Fetterly, Mark Manasse, Marc Najork (2003), "A large-scale Study of the Evolution of Web pages" in the Proceedings of *WWW2003*, May 20-24, 2003, Budapest, Hungary. ACM 1-58113-680-3/03/0005.
2. Fang Li, Huanye Sheng (2003), et al, "The realization of virtual information center for Computational Linguistics" in the proceeding of conference for computer software and application of China. PP.324-329 Shanghai 2003.
3. Gary Simons, Steven Bird (2003), "The open Language Archives Community: An Infrastructure for Distributed Archiving of Language Resources" *in Literary and Linguistic Computing special issue on "New Directions in Humanities Computing".*
4. Joanne Capstick (2002), et al. " COLLATE: Competence Center in Speech and Language Technology" in *Proceedings of the Third International Conference on Language Resources and Evaluation,* Las Palmas, Canary Islands, Spain.2002.
5. Steven Bird, Gary Simons (2003), "Extending Dublin Metadata to Support Description and Discovery of language Resources" *in Computing and the Humanities 37.*
6. Uszkoreit, H (2002), New Chances for Deep Linguistic Processing, In: Proceedings of COLING 2002, Taipei.

# Reduced Human Interaction via Intelligent Machine Adaptation

Hermann Többen, Hermann Krallmann

Technical University of Berlin, Institute Systems Analysis
Franklinstraße 28/29, 10587 Berlin

## Introduction

E-Commerce as one of the predestined enablers of global trade itself is today's subject of increasing competition within the vast globalization of markets. While the latter is still in progress e-commerce steadily seeks out for new unique selling propositions. These attempts converge to the vision of a highly individual preferably one to one customer service with flexible adaptation to his needs. Aside the arrangement of internet sites product advisors have been created in order to navigate the customer through plenty of product and/or services. Examples for those systems are Avatars such as the 'Intelligent Product Advisor', which is based on Intelligent Agent technology and deployed in several web sites (Krallmann 2003). However each request for products and services needs to be executed via an internal business process. The system is supposed to sell automatically only those combinations of products and/or services, which have a priori been considered to be a case for the company's business processes. The flexibility of an e-commerce system thus is mainly restricted by its available business process functionality. This limitation is even more evident, if cross-selling aspects are taken into account. On the base of a buyers' market a company has to be capable of selling also products and services of external vendors. Today's implementations of those features ground on tedious work for setting up the according business processes. Since the markets tend to change not only frequently but also more rapidly the costs for the installation of business processes in particular between different companies represent the critical factor with concern of the pay-off. The combination of the IPA and the CoBPIA system now makes it possible to sustain the business processes in a companies back-end up to a certain extend towards the needs of every single customer.



*G. Hommel and Sheng Huanye (eds.), Human Interaction with Machines,* 85-98.

## Basic System Architecture

In a block diagram the overall system architecture is depicted in figure 1. Three major interfaces are addressed by the identifiers MMI, S1 and S2. The Man-Machine-Interface (MMI) supports the end customer to interact comfortable with the e-commerce system. The Intelligent Product Advisor (IPA) conducts this conversation as a front-end by help of a specific dialogue capability based upon knowledge about products and services. This knowledge is represented in semantic networks and makes it possible to steer the client through the assortment of goods and applicable services. At this point external products and services from different vendors are already incorporated in the discourse of consultancy.

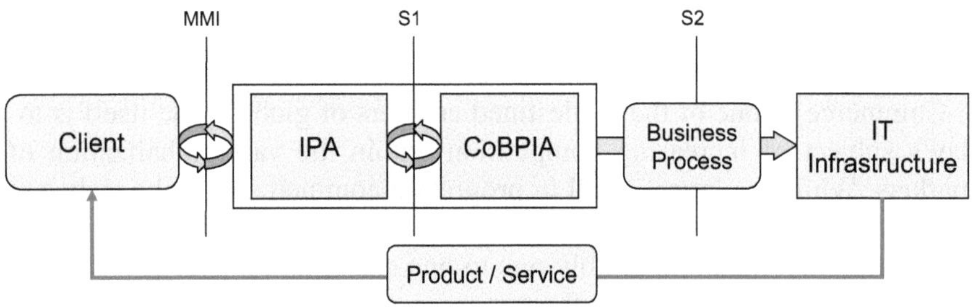


**Fig. 1.** System Architecture.

As a result of this dialogue with the customer a combination of products and services, which has a-priori been checked towards compatibility and feasibility, is exchanged across the S1 interface that ties the IPA together with the CoBPIA process configurator. CoBPIA analyses the request, derives the necessary activities to be performed in order to fulfill the request and finally configures the collaborative business processes, which hence may span across a companie's boundaries. With regard to the latter the configuration process is also highly collaborative and involves the partnering companies. The collaborative business process is passed over to the IT infrastructure via the interface S2. Therefore the internal representation of the business process may be transformed into a target representation like BPEL4WS via a XML transformation. There processes are then installed by means of a workflow-engine like for example the BPEL Process Manager[1] engine. The products and/or services are finally made available to the client via the execution of the business process on the work-flow engine.

[1] Formerly known as Collaxa BPEL4WS engine.

## Automated Consultancy Services

Based upon Newell's hypothesis of emerging intelligence based upon pure symbolic processing various approaches arose within the research domain in order to acquire, store and process knowledge via specific formal systems (Newell 1982). As one of the latter semantic networks represent an organisational model in order to formalize definitions and terms of an arbitrary domain (Quillian 1968). A directed graph is used for visualization, nodes represent subjects or objects of the discourse, meanwhile the edges reflect the relations between these nodes.

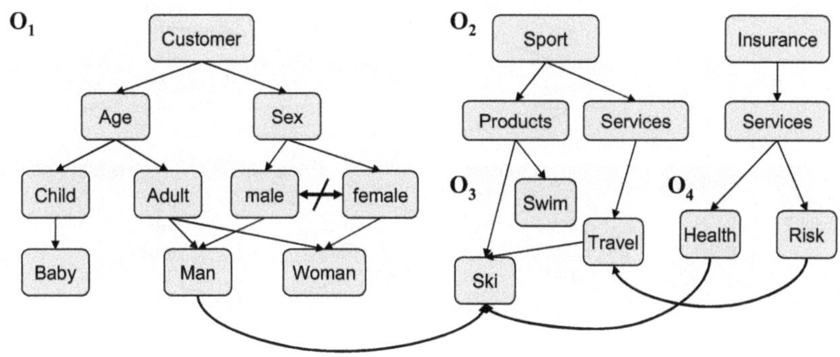


**Fig. 2.** Semantic Network for Knowledge Representation in Ontologies.

The IPA is equipped with such machine processible knowledge about products, services and the customers themselves. The knowledge obeys to a predefined structure also known as ontology. Since this structuring represents a normative grounding for the partnering companies to define their products and services it is possible to link their knowledge up and make it available during the dialogue with the client.

An example semantic network is depicted in figure 2. The customer and his characteristics are described in the ontology $O_1$. He can be discriminated by age and gender. It's necessary to ensure here, that a customer may not be of gender male and female at the same time as indicated by excluding relations. The shared ontology $O_2$ provides concepts for products and services in two different retail areas. In the example sport and insurance are defined. Two further ontologies $O_3$ and $O_4$ contain the products and services of two external retailers one of those selling articles for sport and the other concerned with insurance services. Inside the system each customer is represented by a separate IPA instance. Their mutual interaction is covered by the graphical interface MMI as depicted in figure 1. The IPA guides the customer through an individual, dynamic and menu-oriented counselling (see figure 3). The entire dialog breaks up into dialog activities where each of those is further divided into two elementary dialog activities. In order to generate high-grade recommendations the IPA in addition makes usage of knowledge about the addressed market domain as well as the customer's preferences and his immediate environment. The strategy for the advisory service is thereby implemented through the organization of the knowledge inside this semantic network.

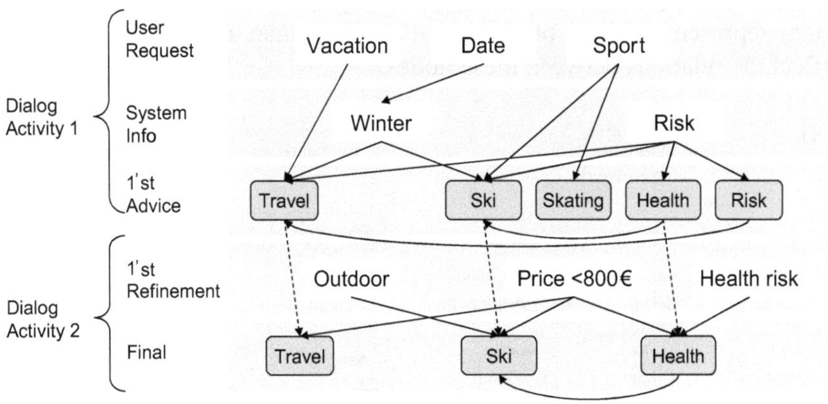


**Fig. 3.** Derivation of Dialog Activities.

In the example the customer asks for vacation at a specific date and is also interested in sports. The system resolves for a travel and – since the proposed date is resolved to be in winter times – for skiing and ice skating as sport activities. In addition to this the IPA may also sell a health-insurance as a cross-selling product, since also a significant correlation exists between skiing and physical injury. After this the first dialog activity is completed and another might start based on further indications from the customer. In the example the customer prioritises the health insurance and constraints, that all products and services should cost less that 800€. As reaction on the customer's feed-back the agent generates new recommendations and starts another dialog activity by offering a refined list to the customer. The latter confirms according to his interests and feeds back to the agent. The number of dialog activities within this process is in principal not limited, but the customer is always permitted to prematurely complete or even to abort the dialog. The navigation in the advisory process, respective the interaction with the customer in the dialog and the creation of recommendation is controlled by the IPA's navigator algorithm. Technically a completed dialog corresponds to a route respective a traversed path in the semantic network.

## Connecting with Business Processes

Actually an arbitrary product basket ordered by the customer may be restricted by capabilities of the underlaying business processes. In the example above is not known in advance, which products may be selected since cross-selling with partnering companies is integrated. Hence the requirements for the business processes diversify with each new selection hence cannot be considered in advance leading to the conclusion, that relevant business processes need to be set up at demand time. For this the IPA is closely coupled with the CoBPIA system via the interface S1. CoBPIA provides means for the transformation of a request formulated as set of constraints into one or more business processes. In a first outline the IPA transforms the assortment of products and services into constraints, which comprise amounts, prices, the providing company as well as certain interdependencies between the items like the binding of the health insurance to the travel for skiing.

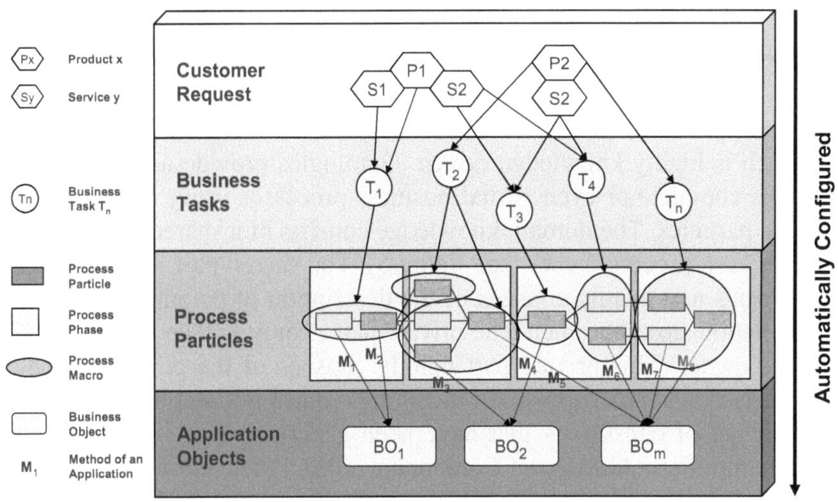


**Fig. 4.** Incremental Derivation of a Business Process.

This request is then handed over to CoBPIA. Figure 4 depicts the layered concept of abstraction towards the construction of business processes. The upper-most layer contains the elements of the shopping basket as selected by the customer. Services may there be applied to products or even other services. Starting with this CoBPIA initially identifies the relevant business tasks, which represent comprehensive activities in order to solve for a well know operational business problem. This might for example be an ordering activity, the delivery of a product, specific furnishing et. al.

The business tasks already indicate which company will take stake in mutual collaboration. Next CoBPIA breaks down the business tasks by configuring a network of elementary activities, which fulfill the preceeded set of constraints, as still assigned with each business task. The elementary activities are represented by process particles, which embody a clear defined functionality as being part of a business process. In addition these particles may be aggregated up to macros. Once an assignment for the constrains has been found, the resulting directed graph of activities represents the business process that maps for the operational execution of the customers request. If no business process can be found, the negative result is reported via the interface S1 to the IPA. The latter then renegotiates with the customer for a modification of his product assortment. A successfully composed process needs then be introduced into the IT infrastructure by mapping elementary activities onto the business object layer via the interface S2. The first outline of CoBPIA contains therefore a BPEL4WS compatible workflow-system Oracle BPEL Business Process Manager.

## Configuration of Business Processes

The principal architecture of CoBPIA is depicted in figure 5. The chosen approach is highly knowledge driven. Ontologies provide a vocabulary to describe complete or even partial business processes using the concept of process particles. The domain knowledge consists of a shared and a private part for each process owner (see figure 5). The shared part is known to all participants and mainly consists of the description of parameter types and their relations to each other. The private part is only known to the specific process owner. This private part mainly consists of the process particles the process owner can apply. An incoming request initiates a search on the possible set of constraints that have been generated from the knowledge base by the 'Particle Request Component' (PRC). The knowledge base itself contains the instantiated process particles and the relevant domain knowledge, which is transferred from the shared ontology. A specific search algorithm, which is provided to compose a valid local process, gets a request, that is formulated using the ontology. Thereby CoBPIA has to handle different types of requests. Either the user defines only the input and output or a fragment of a process can be used as input, which then consists of an incomplete set of process particles. Weather a process particle can be added on a process fragment is checked by the 'Constraint Satisfaction Problem' component (CSP). These checks are accomplished by solving local CSPs. The constraints and variables describe the possible applications of the process particles. The CSP component has two functions on the local level. First it checks if it is possible to add a process particle. Second it verifies if the local process as a whole is a valid process concerning all relevant global constraints.

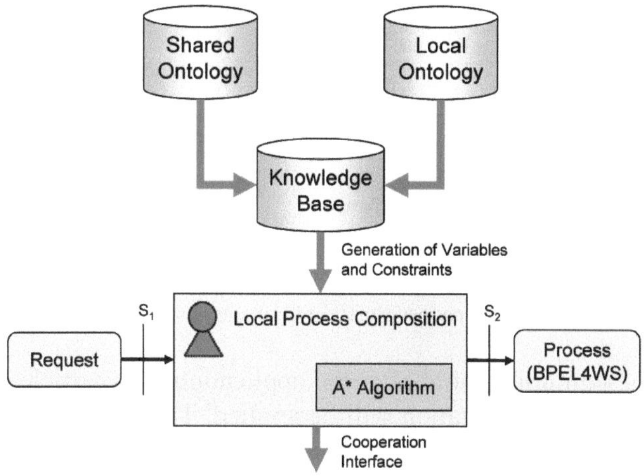


**Fig. 5.** CoBPIA Block Architecture.

The collaboration interface provides means for the interaction between partnering process owners, who also compose local processes at their sites which need to be negotiated with regard to mutual functional verification. For sake of execution on an appropriate platform the created processes are transformed out of the internally used representation into the BPEL4WS format.

## Knowledge Representation

In order to describe the CoBPIA meta-model which enables for the definition of the vocabulary that is needed to build complex business processes the Web Ontology Language (OWL 2005) is deployed. OWL is based on XML Data types and the Resource Definition Framework (RDF 2005) and provides a set of constructs to build ontologies. Figure 6 depicts the most important concept of the CoBPIA meta-model – the *ProcessParticleTemplate* – and its properties. The *ProcessParticleTemplate* is used to describe parts of processes as well as complete processes. In fact processes and process particles are synonyms as the same means are used to describe them. *ProcessParticleTemplates* transform a set of *InputParameterTemplates* to a set of *OutputParameterTemplates*. *ProcessParticleTemplates* can describe pre- and post-conditions concerning their inputs and outputs both using the concept *Constraint*. The preconditions of a *ProcessParticleTemplate* must hold to be able to apply this particle.

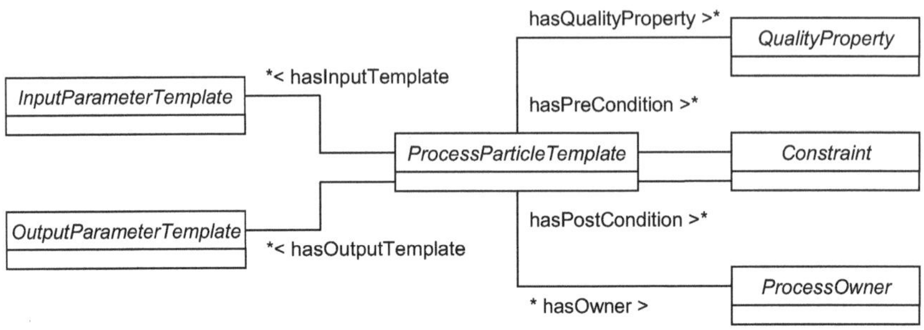


**Fig. 6.** Process Particle Overview.

The post-conditions hold after the application of a particle. For the prototype a string representation will be applied. The *ProcessOwner* attached to the *ProcessParticleTemplate* by *hasOwner* is responsible for the execution of the *ProcessParticle* and possibly provides it to others. Hence the concept *ProcessOwner* is needed to represent each agent within the CoB-PIA agent society unambiguously.

## Modeling the Business Process Control Flow

Figure 7 reflects how the control flow is defined on meta-level. Two parts of the meta-model are needed to be distinguished even both are closely related. The first one describes the outer appearance of processes and process particles. This is called the template representation part. The second part describes the internal structure of processes and process particles. This is called the internal representation part and is emphasized in the figure with gray shades. The concept *ProcessParticleTemplate* describes templates for processes, which maybe used within other processes. The Boolean property *isPublic* is true, if this particle may be taken as a sub particle for composite particles to be generated. Otherwise it may only be used as a sub part of another template which is already present within the ontology. The concept *ProcessParticleTemplate* is not intended to be directly instanced. So far three main sub concepts have identified:

- AtomicTemplate,
- InterfaceTemplate and
- CompositeTemplate.

*AtomicTemplate* and *InterfaceTemplate* are both sub concepts of concept *SimpleTemplate* which means that they can not be decomposed into further sub parts. The *AtomicTemplate* can be compared to what is called 'activity' in other approaches (V.d. Aalst et al. 2003).

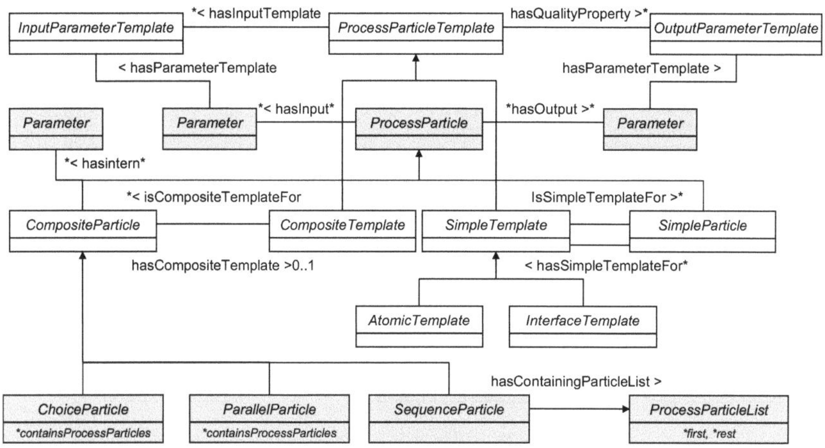


**Fig. 7.** Control Flow Meta-Model.

It can be directly mapped to an activity within an organization, no matter whether it is being executed automatically or performed manually. The *InterfaceTemplate* describes a process particle whose control flow is not known. This concept can be used to describe requests for process composition. The *CompositeTemplate* is used for particles which consist of more than one sub part. Each of these above mentioned concepts of the template representation part has a corresponding concept in the internal representation part. *SimpleParticle* and *CompositeParticle* are the counter pieces of *SimpleParticleTemplate* and *SimpleCompositeTemplate* respectively. Parameters are the internal representation of parameter templates. The *CompositeParticle* is not supposed to be directly instanced. Its specializations define the control flow. The particles contained by a *ParallelParticle* as comprised by the concept *containsProcessParticles* may be executed concurrently. Particles contained by the concept *ChoiceParticle* are alternatives: one of these contained particles whose pre-conditions are met will then be applied. Particles contained by a *SequenceParticle* as comprised by the concept *hasContainingParticleLists* are executed successively. Since *CompositeParticles* may not only hold other particles but the same sub particle even more than once the distinction of a template and an internal representation part is justified here.

## Modeling of Basic Data-Flow

Up to here it was shown how the control flow of processes can be defined using the CoBPIA meta-model. Now the data flow within and between processes needs to be expressed. Since at least two interacting process particles are needed to constitute a data flow, the latter is defined by *SequenceParticles* (*definesDataFlow*) using the *DataFlow* concept (see figure 8).

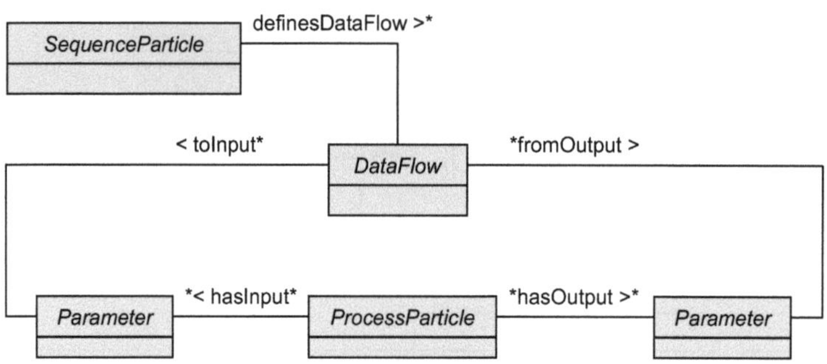


**Fig. 8.** Data Flow Meta-Model.

For example a data flow may be defined between the two *ProcessParticles* A and B if both are directly contained by the same *SequenceParticle* C, B succeeds A, has an input iB, while A has an output oA and finally the types of iB and oA match. Figure 8 shows how parameters of *ProcessParticleTemplates* and their types are defined. There *InputParameterTemplate*, *OutputParameterTemplate* and *InternParameterTemplate* are sub concepts of *ParameterTemplate*. The property *isParameterTemplateOf* points back to the *ProcessParticleTemplate* it is declared for. In fact it is an inverse property of *hasInputTemplate* of *InputParameterTemplate* and *hasOutputTemplate* of *OutputParameterTemplate* respectively. The counter piece of *isParameterTemplateOf* in the internal representation part is *isParameterOf*. Furthermore there are three sub concepts of *InputParameterTemplate*: *ConsumedInputTemplate* describes inputs that are destroyed during the execution of the process particle that 'has' this kind of input. *LockedInputTemplate* describes inputs that are not destroyed during the application of the corresponding process particle. They can serve as inputs for succeeding particles again. *ReferencedInputTemplates* work like *LockedInputTemplate* but the difference is that they can serve as inputs for other concurrent particles. In fact *InputParameterTemplate* is not intended to be instanced directly but its three sub concepts. But which of these three concepts may be applied depends on the type of the parameter. The concept

*ParameterType* builds the root super concept for all parameter type definitions (types). There are two specializations: *ReusableType* and *ShareableType*. All types allow the application of *ConsumedInputTemplate* when the specialization of an *InputParameterTemplate* for a *ParticleTemplate* is chosen. If the defined type is a sub concept of *ReusableTypeTemplate* then *LockedInputTemplate* is allowed as well. Finally, if the type is a sub concept of *ShareableType* then *ReferencedInput* may be applied. Types are also defined as part of the domain knowledge.

### Process Composition

The Process Composing Component (PCC) does the basic work to compose processes. Therefore it uses process particles as defined in the knowledge base and further more utilizes the CSP component to validate the composed processes. This initial request contains several information like the input for this process and furthermore the final output, which the process should create. These can be choosen from a list containing the known parameter types in the ontology. The major challenge for the composer is that no information about the process itself is provided. This confinement is caused by the fact that an automatic composition has to be done, what implicates that the user just defines the goal but not the process itself in terms of activities, control and data flow.

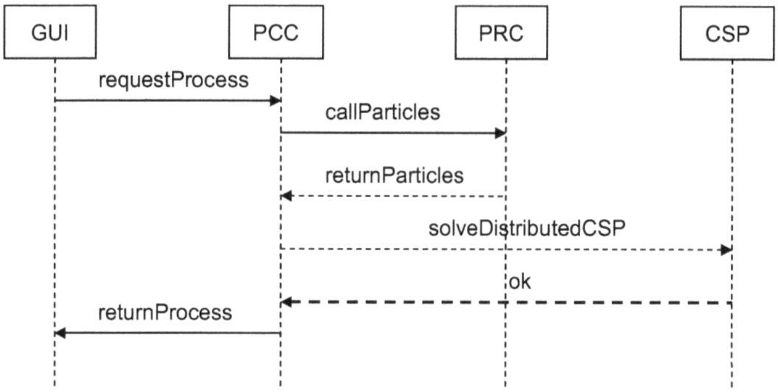


**Fig. 9.** Sequence Chart for Distributed Process Creation.

As depicted in figure 9 the initial request starts the creation of a new process. From this point on an iterative composition process starts, which creates different process alternatives, which are successively elongated with new process particles. The first step is to receive fitting particles from the *ParticleRequester* component as attached to the knowledge base. Here the *ParticleRequester* retrieves for fitting particles and inferences also on subclasses. Next the PCC creates the possible valid process combinations.

At this point initial inputs are matched with the inputs for the actually deployed particles. The parameter types of every combination fit together, so that the first part of the condition for correctness is satisfied. Now the constraints defined on the parameters have to be validated. This is done for each combination by a local constraint solver. If no inconsistencies exist between the constraints the combination will be put into the internal search tree. This whole process is iterated until a process is found where the final output matches the predefined goal output. If this holds, the distributed CSP is started. This component solves the more complex distributed constraint problem. If this constraint problem has also been successfully solved the process can be delivered back. If not, the composing engine starts over to find the next valid process, which could fulfill the goal.

### *Building Process Combinations*

The combination building has to find all valid variations to sequence the particles. It is possible, that particles can be grouped together in different ways each of them being a valid combination. The example depicted in figure 10 illustrates how the combination building.

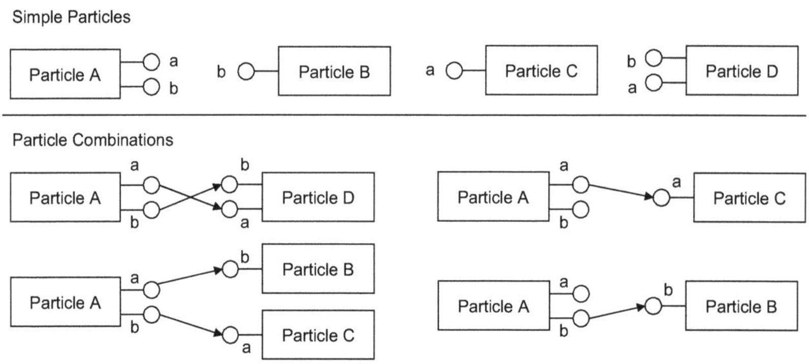


**Fig. 10.** Example for Process Combination Building.

Particle A offers the output of the previous iteration. Therefore the *ParticleRequester* is called with the parameters a and b and returns the particles B, C and D. The combination builder creates all four combinations. The algorithm works recursively through the set of particles and checks which parameter types fit together, because the next step is to assure that all input parameters of the connected particles are used. For instance particle D has to be matched on both input parameters. The algorithm assures, that all possible matches of parameters will be found, but in an unsorted way. If the parameter types of particle A would be the same, there will only two combinations. Every match of a parameter is stored in an object called *ParameterMatch*. This object contains a 'from' and a 'to field, where the parameters are stored. Every combination consists of a set of particles a *ParameterMatches*. The match itself is created by comparing the *typeinfo* field of the parameter. When the combination builder has completed it is assured, that the combinations could be valid process elongations.

## Related Work

Automatic process configuration can be seen as a special approach for flexible business processes. Peculiar in the field of 'Workflow Management Systems' (WfMS) many projects deal with issues of flexibility, where overall quality of the processes shall be improved and run-times reduced (Heilmann 1994). Nonetheless in practice WfMSs often freeze processes instead of introducing more flexibility since they are not able to meet the requirements resulting in a highly dynamic environment (Maurer 1996). Agent-based approaches like 'Advanced Decision Environment for Process Tasks' (ADEPT) introduce more process flexibility. There agents represent carriers of services, that may be mutually recombined during run-time (Jennings et al. 1996). Similar to ADEPT the approach 'Agent Oriented Organisations' (AEGIS) also relies on Agents, that represent organisational units within a company and act as "intelligent, autonomous problem solvers, having their own aims, resources, capabilities and beliefs" (Unland et al. 1995). Shepherdson et al. describe the disadvantages of Intelligent Agent approaches like ADEPT and AEGIS and criticises aside the intransparency of the business processes also the high costs for implementation and reengineering of agent-based workflow systems (Sheperdson et al. 1999). A more comprehensive summarization and comparison of adaptive and especially agent-based work-flow management systems for business process execution can be found in (Lindner 2003).

## References

Heilmann H. (1994) Workflow Management: Integration von Organisation und Informationsverarbeitung, HMD 176/1994, pp. 8-21.

Jennings et al. (1996) ADEPT: Managing Business Processes using Intelligent Agents. In: Proceedings of the 16th Annual Conference of the British Computer Society Specialist Group on Expert Systems (ISIP Track). Cambridge, UK, 1996, pp. 5-23.

Jennings et al. (2000) Implementing a Business Process Management System using ADEPT: A Real-World Case Study. In: Int. Journal of Applied Artificial Intelligence 14 2000, Nr. 5, pp. 421-463.

Krallmann H. (2003) Intelligente Produktberatung im Internet. In: Wolfgang Kersten (ed) E-Collaboration Prozessoptimierung in der Wertschöpfungskette, Gabler Edition Wissenschaft (HAB), Wiesbaden, pp. 111-130.

Lindner I. (2003) Flexibilität von computerunterstützten Geschäftsprozessen - Kategorisierung von Forschungsansätzen und Erarbeitung von Validierungskriterien, Diplomarbeit, Berlin.

Maurer G. (1996) Von der Prozeßorientierung zum Workflow Management. Teil 2: Prozeßmanagement, Workflow Management, Workflow-Management-Systeme. Arbeitspapiere WI - Nr. 10/1996. Lehrstuhl für Allgemeine BWL und Wirtschaftsinformatik. Universität Mainz. Mainz 1996.

Newell A. (1982) The Knowledge Level. In: Artificial Intelligence, Vol. 18, pp. 87-127.

OWL-S Version 1.1 (2004), to be found unter http://www.daml.org/services/owl-s/1.1/ (Last accessed 02/23/05).

Quillian M. Ross (1968) Semantic Memory. In: Semantic Information Processing, hrsg. Minsky, M., MIT Press, Cambridge, MA, USA, pp. 227-270.

RDF Version 1.0 (2004), to be found unter http://www.w3.org/RDF/#specs, (Last accessed 02/23/05).

Shepherdson J.W., Thompson S.G., Odgers B.R. (1999) Cross Organisational Workflow Co-ordinated by Software Agents. In Ludwig, H. (ed.), Proceedings of WACC Workshop on Cross-Organizational Workflow Management and Co-Ordination, San Francisco.

Unland R., Kirn S., Wanka U., O'Hare G.M.P., Abbas S. (1995) AEGIS: Agent Oriented Organizations; Accounting, Management And Information Technologies, Pergamon 1995, pp. 139-163.

Van der Aalst W.P., Hofstede A.H., Kiepuszewski B, Barros A.P. (2003) Workflow patterns.

# A Novel Watershed Method Using Reslice and Resample Image

Shengcai Peng, Lixu Gu

Computer Science, Shanghai Jiao Tong University, 1954 Huashan Road, Shanghai, P.R.China 200030
tsai@sjtu.edu.cn, gu-lx@cs.sjtu.edu.cn

**Abstract.** A novel Watershed segmentation approach is proposed in this paper. Firstly a review of medical image segmentation was presented where the advantages and drawbacks of the traditional watershed method are discussed. To resolve the existing problem, the proposed algorithm employs a threshold operation according to the range determined by the region of interest (ROI) and the modality of the medical image. And a "reslice" operation was employed to shrink the intensity of the image into a smaller range. Finally a resample operation is introduced to reduce the resolution of the image and this make watershed transform run in shorter time and less memory. The results indicated that the proposed method is efficient in solving the over segmentation problem and faster than classic watershed.

**Keyword.** segmentation, watershed, resample, reslice

## 1 Introduction

Image segmentation is one of the most critical tasks in automatic image analysis. For example, segmentation is a prerequisite for quantification of morphological disease manifestations and for radiation treatment planning, for construction of anatomical models, for definitions of flight paths in virtual endoscope, for content-based retrieval by structure, and for volume visualization of individual objects. The application that provided the incentive for our work was manifestations of disease from CT/MRI data.

Many different algorithms have been proposed to address the segmentation, which may fall to two major streams: model based and region-based methods. The model based method such as Snake or Level set is relatively faster but sometimes can't achieve satisfactory accuracy, especially at the narrow sharp boundary area. The region based method like Watershed [1],[2] or Morphological operations may get more accurate results but more costly. However, Watershed can be parallelized easily to improve its performance, and has been widely used in medical image segmentation.



*G. Hommel and Sheng Huanye (eds.), Human Interaction with Machines,* 99-106.

Since S. Beucher and F.Mever firstly proposed Watershed algorithm [6], many scientists have addressed their efforts to improve it [3]. Some researchers employed diffusion algorithms before running the watershed transform because watershed is very sensitive to noises [8]. Some researchers use probability theory to compute the landscape of the image or use different strategy to compute the saliency of merge to merge the watershed images [7]. But the over-segmentation problem has still not been effectively resolved.

In this paper the traditional watershed was introduced (see Section 2.1) at first. Then it discusses an improved watershed method using reslice image to resist over-segmentation problem and resample image to lower the cost from traditional watershed algorithm (see Section 2.2). The method is especially designed for medical image analysis by using the pre-knowledge of medical imaging phenomenon. For the sake of clarity, a comparison between classical watershed and our method is also presented (See Section 3). A conclusion about our method was drawn at the end of this paper (See Section 4).

# 2 Methods

## 2.1 Traditional Watershed

A widely used implementation of traditional watershed is a method by simulating immersion [4] where an image is regarded as a (topographic) surface. Suppose this surface was immersed into a lake, the water will progressively fill up the different catchment basins of the image. Then, at each pixel where the water coming from two different minima would merge, we build a "dam". At the end of this immersion procedure, each minimum is completely surrounded by dams, which delimit its associated catchment basin. The whole set of dams which has been built thus provides a tessellation of the image in its different catchment basins and these dams correspond to the watersheds of our image. (See Fig. 1).

This method was quite efficient but without thinking about the over-segmentation problem. Over-segmentation is a quite common problem as watershed is very sensitive to noise and gives segmentation even to low contrast region, which may belong to the same object. We will employ traditional watershed as a base of the proposed method.

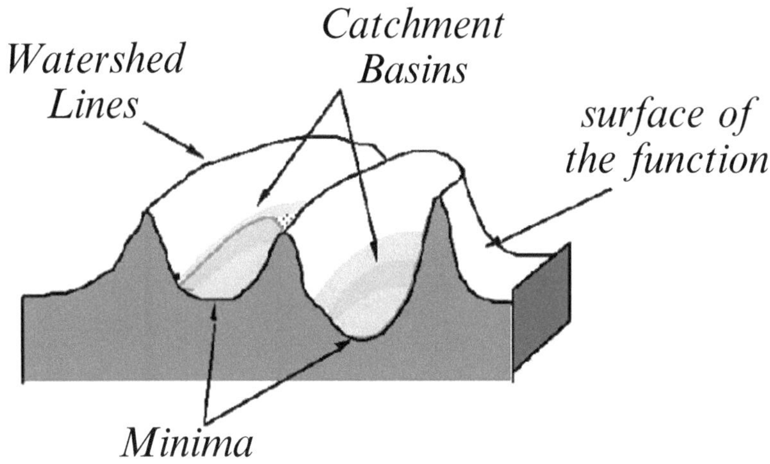


**Fig. 1.** Minima, catchment basins and watersheds.

## 2.2 Watershed Using Reslice and Resample Image

There are three preprocessing steps in our method. Firstly the threshold operation is applied according to the range determined by the region of interest (ROI) and the modality of the medical image. As we know, different organs are presented in different intensity ranges in a medical image because of their different physical features (Fig. 2). After the modality of the input medical image data was defined, the intensity range of a specified organ can also be roughly determined. Based on this pre-knowledge, the regions with the intensity out of this range can be ignored. The processing makes the image simpler and reduces the complexity of the whole procedure.

Secondly we make a reslice of the image. Here reslice means presenting the image with a smaller intensity range. For example, if we use one byte to present one pixel, as one byte consists of 8 bits, each pixel has one of the $2^8$=256 grey levels. Usually a medical image uses four bytes for every pixel, but after the first step of our method, only the tissue with the intensity in a specified range was left, so it is possible to present the image with shorter bytes. And more importantly, this operation merges the pixels with small differences together, and the over segmentation problem can be resisted. The original image and the resliced image are shown in Fig. 3. It must be emphasized that we can only reduce the grey levels to a proper number; otherwise the detail of the image may be lost.

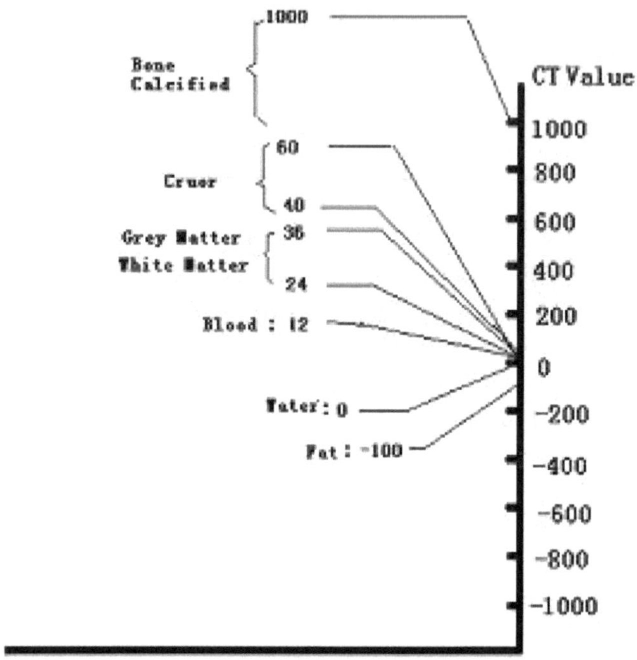


**Fig. 2.** Different organs are presented in different intensity ranges in a CT image.

Finally, the preprocessed image is sent to a resample procedure. The resample produces an output with different spacing (and extent) than the input. Linear interpolation can be used to resample the data. Here the resample is used to reduce the resolution of the image, so that the watershed transform requires less memory and run faster. Furthermore, the resample operation helps to resist the over segmentation problem because the small regions are ignored and reduced during this operation. A comparison of a brain image and its resample image (In every dimension the resample rate is 0.8) are shown in Fig. 4.

## 3 Results and Discussions

We implemented our experiment in a normal PC with Pentium 2.4G CPU and 1G DDR RAM, where the program was written in Python + VTK environment and run on a Windows platform. Two datasets were used in this experiment: One is a brain MRI data [5] downloaded from BrainWeb[1]. It is a 181*217*181 voxel volume, with isotropic 1 mm3 voxels. Another one is a real clinical canine cardiac CT data, which is a 512*512*86 voxel volume with isotropic 0.35*0.35*1.25 mm3 voxels.

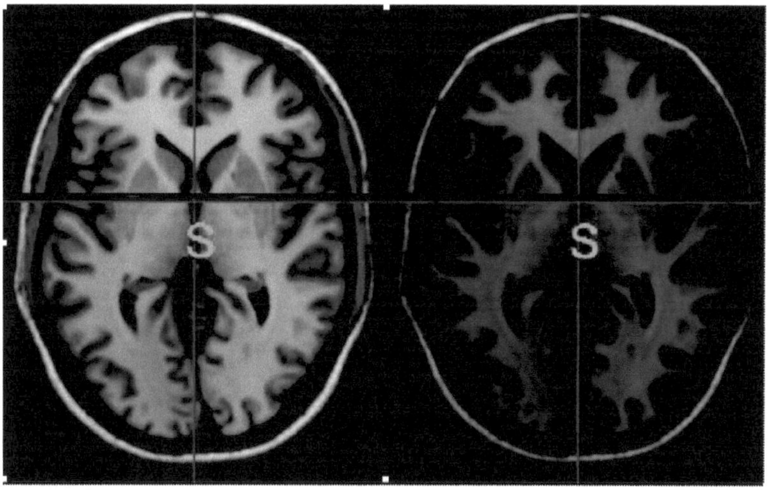


**Fig. 3.** (Left) The source image. (Right) The image after reslices operation.

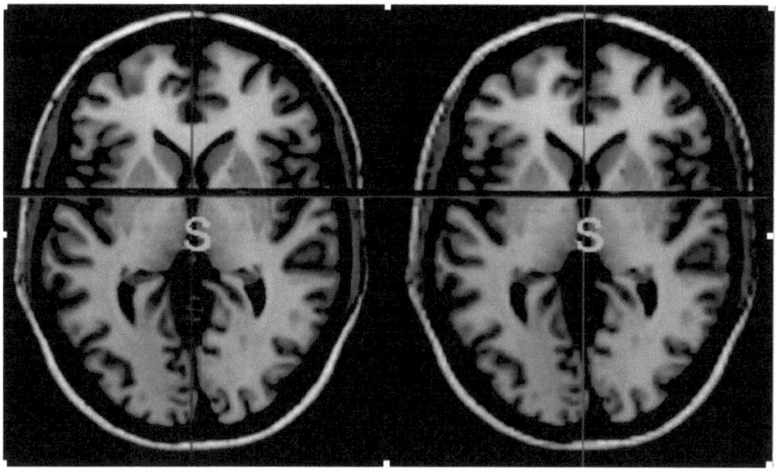


**Fig. 4.** (Left) A brain image. (Right) The resample image (with resample sample rate 0.8 in every dimension)·

Three comparisons to demonstrate the effect of every step of our method was conducted. Fig. 5 indicates that when the threshold operation is

---

[1] http://www.bic.mni.mcgill.ca/brainweb/

applied (Fig. 5(b)) at first, the regions out of the brain are resolved into background. Fig.6 shows the results of the watershed transform applied to original image (a), reslice (b) and resample (c) images, respectively. The amount of the watershed regions is decreased apparently. From Fig. 5, we also found that some part of the ROI is merged to background unexpectedly. This is often caused by connections between the ROI and the surrounding structures. For better results, the morphological opening operation was employed to deal with this problem by breaking the connections.

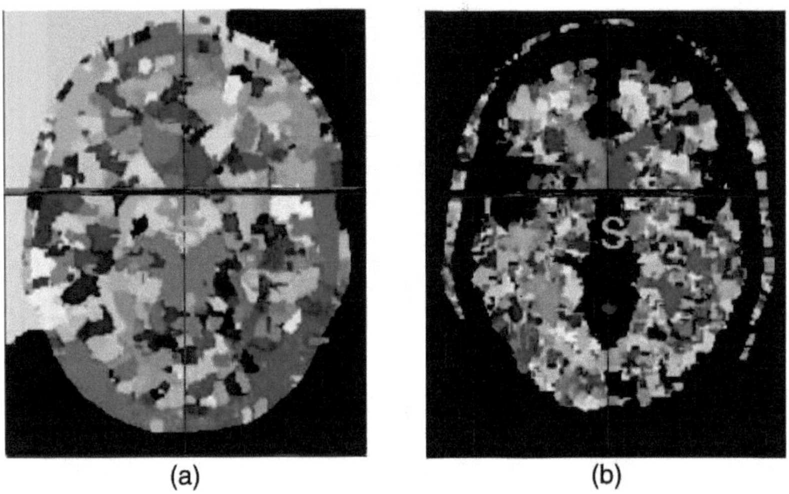


**Fig. 5.** Comparison between the normal Watershed and the Watershed on the threshold image (a) Normal Watershed (b) Watershed with threshold operation.

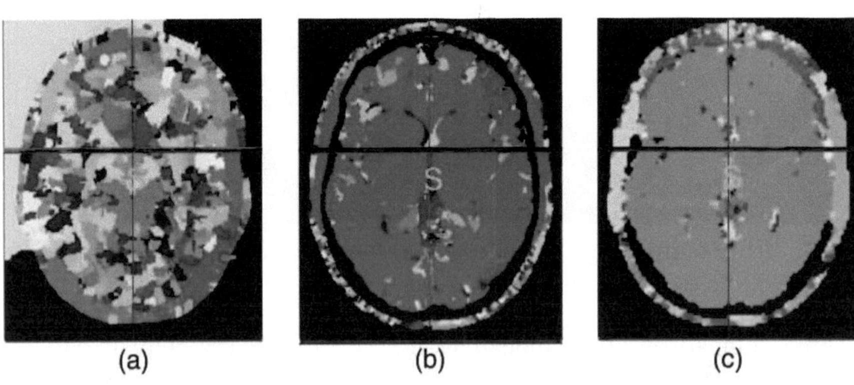


**Fig. 6.** The results of the watershed transform applied to original image (a), reslice (b) and resample (c) images, respectively.

The comparison reveals that proposed approach can significantly resist the over-segmentation problem. The average amount of watershed regions in ROI is described as Table 1:

- Classical watershed generated 244 regions (Fig. 6(a)) of ROI;
- Watershed using reslice image generated 84 regions (Fig. 6(b)) of ROI;
- Watershed using reslice and resample image generated 32 regions (Fig. 6(c)) of ROI.

The efficiency of the proposed method has also been evaluated with the classical one. The proposed watershed cost 95s to process the brain data, when the classical Watershed required 90s.

**Table 1.** Comparison between the traditional watershed and our method in generated areas and time consuming.

| | Generated regions of ROI | Time consuming |
|---|---|---|
| Traditional Watershed (Fig. 6(a)) | 244 | 90s |
| Watershed using reslice image (Fig. 6(b)) | 84 | 95s |
| Watershed using reslice and resample image (Fig. 6(c)) | 32 | 95s |

## 4 Conclusion

In this paper, an improved watershed method using reslice and resample image was proposed. The reslice operation significantly decreased the amount of the watershed regions in ROI to 34%. And resample operation can further decrease this amount to 38%. It means that the over-segmentation problem was resisted efficiently. But the computing cost has not been improved. To save the cost, it is recommended to implement the method with parallel strategy in the future.

## References

[1] J.B.T.M. Roerdink, A. Meijster, "The Watershed Transform:Definitions, Algorithms and Parallelization Strategies", Fundamenta Informaticae, vol. 41, pp.187-228, 2000.

[2] Kari Saarinen, "Color Image Segmentation By A Watershed Algorithm And Region Adjacency Graph Processing", IEEE International Conference Image Processing, Proceedings, ICIP-94, pp.1021-1025, 1994.

[3] L.Shafarenko, M.Petrou, J.Kittler, "Histogram-Based Segmentation in a Perceptually Uniform Color Space", IEEE Transactions On Image Processing, vol. 7. No.9, 1998.

[4] Luc Vincent, Pierre Soille, "Watersheds in Digital Spaces: An Efficient Algorithm Based on Immersion Simulations", IEEE Transactions On Pattern Analysis And Machine Intelligence, vol. 13, No.6, 1991.

[5] R.K.S. Kwan, A.C. Evans, G.B. Pike, "MRI simulation-based evaluation of image processing and classification methods", IEEE Trans. Med. Imag., vol.18-11, pp.1085-1097, Nov. 1999.

[6] S.Beucher, F.Meyer, "The morphological approach to segmentation: The watershed transformation", Mathematical Morphology in Image Processing, New York, pp. 443-481, 1993.

[7] V.Grau, A.U.J. Mewes, M. Alcaniz, "Improved Watershed Transform for Medical Image Segmentation Using Prior Information", IEEE Transactions On Medical Imaging, vol. 23, No. 4, 2004.

[8] Zhao Jianwei, Wang Peng, Liu Chongqing, "Watershed Image Segmentation Based on Wavelet Transform", Acta Photonica Sinica vol. 32, No. 5, China, 2003.

# Stable Motion Patterns Generation and Control for an Exoskeleton-Robot

Konstantin Kondak, Günter Hommel

**Abstract.** The presented algorithm enables an online generation of stable motion patterns with minimal computational effort. The motion patterns are specified by the user with few task parameters such as: step size, step duration and walking speed, which can be changed in the generation process or during the motion. The algorithm is based on a feedback loop scheme, composed of a linearized and decoupled non-linear model of the biped. To make the algorithm insensitive to model parameters uncertainty a sliding mode controller was integrated in the control loop. The balancing/stabilization of the robot is achieved by appropriate change of the joint accelerations. The performance of the algorithm was studied and demonstrated in a simulation. The algorithm can be used as motion patterns generator, as a high level controller as well as for balancing/stabilization of existing movement trajectories.

## Introduction to the Problem

The motion control system for an exoskeleton is composed of two main blocks: recognition of the human motion intention and control of the motion. This work is devoted to the second block which is composed of the generation of motion patterns and motion control. We consider the motion of the whole body, so that the keeping of balance on one foot or two feet becomes an important issue. Due to the huge amount of possible motion patterns which can be initiated by a human within the exoskeleton the usage of the precomputed stable trajectories is quite difficult. Therefore the algorithm should generate stable motion patterns, corresponding to recognized motion intention of the human, online. The human within the exoskeleton will be approximated with a model composed of several rigid bodies connected by means of joints, so that the whole system can be considered as a biped robot.

Despite the existence of advanced biped robots e. g. [1], the problem of the stable or balanced motion generation remains unsolved due to a fundamental difficulty which the authors would like to point out before starting the description of the proposed algorithm.

The 3D model of a biped composed of the trunk, two upper, two lower legs and two feet, will be considered. Two hip joints are modeled as spherical joints and two ankle joints as universal joints. Each of two knee



*G. Hommel and Sheng Huanye (eds.), Human Interaction with Machines,* 107-116.

joints has one rotational axis. The vertical z-axis and horizontal x-axis lie in the sagittal plane. The dynamical equations which describe the movement of a biped have the following form:

$$\mathbf{M}(\mathbf{q})\dot{\boldsymbol{\omega}} = \mathbf{f}(\mathbf{q},\boldsymbol{\omega}) + \mathbf{T} \tag{1}$$

where $\mathbf{q} = (q_1,\ldots,q_{12})$ is the vector of generalized coordinates, which are the joint angles, and $\boldsymbol{\omega} = (\omega_1,\ldots,\omega_{12})$ is the vector of corresponding generalized velocities. The matrix function $\mathbf{M}(\mathbf{q})$ takes into account the mass distribution, and the vector function $\mathbf{f}(\mathbf{q},\boldsymbol{\omega})$ describes the influence of both inertial forces and gravity. The elements of the vector $\mathbf{T}$ are the torques in the joints. The dot denotes the time derivative in a Newtonian reference frame.

The kinematic equations for the model are obviously:

$$\dot{\mathbf{q}} = \boldsymbol{\omega} \tag{2}$$

If the foot of the robot would be rigidly attached to the floor, the problem of motion pattern generation became trivial (practical implications such as joint angles and torque limits as well as knowledge of model parameters not considered). For each given trajectory of coordinate vector $\mathbf{q}$, the torques $\mathbf{T}$ which realize this trajectory could be easily calculated from eq. (1). Unfortunately, in practical applications the robot foot is not attached to the floor and can react on it only vertically downwards and tangential to the floor surface. Here the possible rotation of the robot around the foot edge should be considered. To account for this, the concept of zero moment point (ZMP), which was proposed by Vukobratovic almost 30 years ago [2], has been used. The position of ZMP, denoted as $\mathbf{r}_{zmp}$, can be expressed in the following form:

$$\mathbf{r}_{zmp} = \boldsymbol{\alpha}(\mathbf{q},\boldsymbol{\omega}) + \dot{\boldsymbol{\omega}}^T\boldsymbol{\beta}(\mathbf{q}) \tag{3}$$

where $\boldsymbol{\alpha}(\cdot)$ and $\boldsymbol{\beta}(\cdot)$ are non-linear functions. Specification of the ZMP movement, i.e. defining the trajectory or position $\mathbf{r}_{zmp}$, implies a strong restriction on possible trajectories $\mathbf{q}$ in form of eq. (3). That means that stable motion pattern generation requires the solution of a boundary value problem (BVP), given by eq. (3) and boundary values for elements of $\mathbf{q}$ and/or $\boldsymbol{\omega}$ at the beginning and at the end of the motion.

Even for a simple model of the biped, composed of a planar non point mass trunk body with massless legs moving in sagittal plane, the solution of this BVP is very complicated and can be computed only numerically.

These two general facts: stable motion pattern generation implies the solution of a BVP and the difficulties in solving this BVP, are the reason for a huge number of publications where different simplifications are made to be able to compute online a stable gait for 3D models.

In the presented algorithm instead of specifying ZMP trajectory or position the allowed area for ZMP is specified:

$$\mathbf{r}_{\min} \le \mathbf{r}_{zmp} \le \mathbf{r}_{\max} \tag{4}$$

This vector ineq. is composed of two components for $r_{zmp\,x}$ and $r_{zmp\,y}$. The usage of ineq. (4) instead of prescribed trajectory or position for ZMP has the advantage that the complicated differential eq. (3) does not have to be satisfied exactly. If ZMP is somewhere within the allowed region, the particular value of $\mathbf{r}_{zmp}$ is not important for the stability.

The rotation around the edge of the foot or violation of condition (4) does not necessarily mean that the biped is going to collapse. Theoretically, this rotation could still be controlled by means of inertial forces generated by movements of the body parts, or stopped by reconfiguration of the kinematic structure of the system, e.g. additional support by the other foot. But the question is whether such a movement has any advantages. In any case, the control of the system during rotation around the edge of the foot requires more effort and is very sensible to disturbances and parameter uncertainties. This kind of movement will be not considered here.
Further, the movement will be called stable or balanced if the condition (4) is satisfied. For most applications the rotation of the robot around the vertical z-axis is automatically prevented by friction forces and is not considered.

## General Description of the Algorithm

In the proposed algorithm generation of motion patterns and control are collected in one single feedback loop. This algorithm can be considered as a global controller that guides the system from some current state $(\mathbf{q},\boldsymbol{\omega})$ to a desired state $(\mathbf{q}_1,\boldsymbol{\omega}_1)$. For task specification kinematic parameters in Cartesian 3D space like foot position, trunk orientation etc. are easier to use compared to the generalized coordinates $\mathbf{q}$ and velocities $\boldsymbol{\omega}$. For the considered system twelve kinematic parameters $x_1,\dots x_{12}$ and twelve corresponding velocities $v_1,\dots v_{12}$ should be chosen. By means of kinematic eqs. $x_1,\dots x_{12}$ can be recomputed in $\mathbf{q}$, $v_1,\dots v_{12}$ in $\boldsymbol{\omega}$ and vice versa. The choice of kinematic parameters is not unique and should be made in order to simplify the task specification of the system.

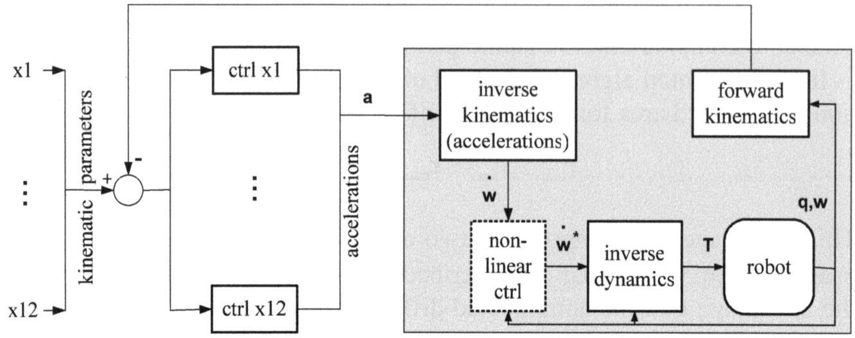


**Fig. 1.** Scheme of the algorithm.

The scheme of the algorithm is shown in Fig. 1. The block *inverse kinematics (accelerations)* computes the generalized accelerations $\dot{\boldsymbol{\omega}}$ from the accelerations of the kinematic gait parameters. For that, a system of linear equations should be solved, which is obtained by double differentiation of the forward kinematics equations. The solution can be performed symbolically, therefore, the formulas for calculation of the elements of $\dot{\boldsymbol{\omega}}$ are available.

The block *robot* represents the real biped or its non-linear model.

The block *inverse dynamics* computes the joint torques $\mathbf{T}$ from the given acceleration vector $\dot{\boldsymbol{\omega}}$ and the current system state $(\mathbf{q},\boldsymbol{\omega})$ using dynamical eq. (1). This block linearizes and decouples the non-linear system which describes the biped, so that the blocks *robot* and *inverse dynamics* are equivalent to the twelve independent double integrators, one for each joint.

The block *forward kinematics* computes the current values for kinematic parameters and their velocities from the current system state $(\mathbf{q},\boldsymbol{\omega})$. For a moment, we assume that the dotted block *non-linear ctrl* has a transfer function equal to 1 and $\dot{\boldsymbol{\omega}} = \dot{\boldsymbol{\omega}}^*$. Due to this assumption, the shaded area in Fig. 1 can be considered as twelve independent double integrators, which describe the behavior of each of the kinematic parameters $x_1,\ldots x_{12}$. These parameters are independently controlled by the blocks *ctrl.*

Until now, the control scheme looks similar to the widely used approach for decoupling, linearization and control for non-linear mechanical systems, especially arm manipulators (see e.g. [3]). The application of the described scheme for generation of stable motion patterns or global control of a biped involves the satisfaction of the stability condition (4). For achieving the stability, the concept of ZMP is used. The stability is achieved by modifying the acceleration vector $\dot{\boldsymbol{\omega}}$ in the block *non-linear ctrl*. This modification will be performed by projecting the vector $\dot{\boldsymbol{\omega}}$ onto the plane, which is denoted by the authors as ZMP-plane. The projection operator requires the solution of a system of linear equations with the dimension less than the dimension of the vector $\dot{\boldsymbol{\omega}}$. This means that computations in the block non-linear ctrl are not expensive and the method can be easily applied to models with more DOF.

## Stability Condition

We consider eq. (3). At each moment, the acceleration of the system $\dot{\boldsymbol{\omega}}$ can be changed arbitrarily by application of corresponding torques $\mathbf{T}$ in the joints (according to eq. (1)). By contrast, the changes in velocities $\boldsymbol{\omega}$ correspond to actual accelerations, and the changes in coordinates $\mathbf{q}$ correspond to actual velocities (double integrator behavior). This allows the coefficients $\boldsymbol{\alpha}(\cdot)$ and $\boldsymbol{\beta}(\cdot)$ at each moment to be considered as constants and the accelerations $\dot{\boldsymbol{\omega}}$ as variables with arbitrary values. Therefore, for each value of $r_{zmp\,x}$ or $r_{zmp\,y}$ eq. (3) describes a plane in the acceleration space at each moment (for each component of $\mathbf{r}_{zmp}$). This plane is called the *ZMP-plane*. Of course, the position and orientation of this plane are changing in time. The ZMP-plane, e.g. for $r_{zmp\,x}$, has a clear physical meaning: choosing the actual acceleration $\dot{\boldsymbol{\omega}}$ belonging to it, we guarantee that the system moves at this moment of time in such a way that the x-coordinate of ZMP will be equal to $r_{zmp\,x}$ (the same also for $r_{zmp\,y}$).

The block *non-linear ctrl* which ensures the satisfaction of the stability condition (4) is implemented therefore as the following algorithm:

1. **calculate** $r_{zmp\,x}$ **corresponding to** $\dot{\boldsymbol{\omega}}$
2. **if** $r_{zmp\,x} < r_{\min x}$ **or** $r_{zmp\,x} > r_{\max x}$

   **then** $r_{zmp\,x} = r_{\min x}$ **or** $r_{zmp\,x} = r_{\max x}$ ; $\dot{\boldsymbol{\omega}}^* = projection(\dot{\boldsymbol{\omega}})$

   **else** $\dot{\boldsymbol{\omega}}^* = \dot{\boldsymbol{\omega}}$

The same also for $r_{zmp\,y}$. The operator $projection(\cdot)$ in step 2 can be implemented in different ways. In our simulations the orthogonal projection yields the best results. More information about the operator $projection(\cdot)$ is presented in [4], [5].

Using the active operator *projection*(·) the geometrical parameters do not behave like double integrators and the motion generated by blocks *ctrl* is disturbed. The operator *projection*(·) should try to redistribute this disturbance over all dimensions of the acceleration space, so that each particular coordinate has only a small disturbance which can be corrected when the operator is not active.

## Periodical Motion

Walking with single support phase will be considered as an example for periodical motion. The following kinematic parameters were chosen for specification of the system state at the end of each step: pelvis position, trunk orientation, swing foot position and swing foot orientation. These kinematic parameters and their velocities are not independent. By means of a simple biped model composed of the mass point and massless legs the authors have shown [6], that following dependences for the end of each step should be considered:

$$\upsilon_x = \frac{L_x}{2}\sqrt{\frac{g}{h}}\coth\left(\frac{T}{2}\sqrt{\frac{g}{h}}\right), \quad \upsilon_y = L_y\sqrt{\frac{g}{h}}\coth\left(\frac{T}{2}\sqrt{\frac{g}{h}}\right)^{-1} \tag{5}$$

Here $L_x$ and $L_y$ are positions of the swing foot; $\upsilon_x$ and $\upsilon_y$ are the velocities of the pelvis; $T$ is the step duration; $h$ is the pelvis height and $g$ is the gravity constant. All kinematic parameters are derived from few task parameters (with respect to eq. (5)): step size, walking speed, pelvis height. These task parameters are used as input to the algorithm.

To produce a periodic gait, all of the blocks *ctrl* in Fig. 1 should force the kinematic parameters to reach their end values at the same time. As described below, each of the kinematic parameters behaves like a double integrator. The following control law can be used to guide a double integrator to the given state in the specified time:

$$a(\Delta t) = \frac{6(x_1 - x) - 2\Delta t(\upsilon_1 + 2\upsilon)}{\Delta t^2} \tag{6}$$

Here $(x,\upsilon)$ are the current and $(x_1,\upsilon_1)$ the desired coordinate and velocity; $\Delta t$ is the remaining time. The description of this controller can be found in [6].

To demonstrate the performance of the algorithm, the following simulation experiment was performed: The biped was standing in some arbitrarily chosen initial state (the center of mass (CM) was within a supported polygon, all generalized velocities were equal to 0). The biped had to start walking with the following task parameters: step size 0.6 m, step duration 0.7 s (walking speed 0.857 m/s). After five steps or 3.5 s the step duration was abruptly decreased to 0.5 s (new walking speed 1.2 m/s) and the biped had to continue walking with the new walking speed. The limits in stability condition (4) were set quite restrictively to: $\mathbf{r}_{min} = (-0.05, -0.05)$ and $\mathbf{r}_{max} = (0.2, 0.05)$. The parameters of the model were chosen similar to those of the human. In Fig. 2 the movement stability characteristics, motion of the ZMP and CM, are shown.

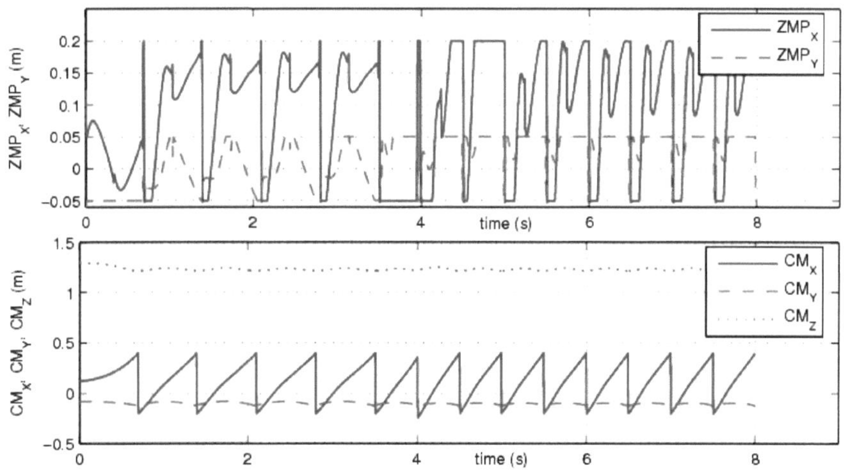


**Fig. 2.** ZMP (top) and CM (bottom) position during the motion.

The coordinates in Fig. 2 are expressed in the reference frame the origin of which is on the floor under the ankle joint of the stance foot (after changing the stance foot the origin is moved to the new position). As one can see, the ZMP remains within the defined region at each time during the movement. At the beginning of the movement and after switching to the new walking speed the patterns for ZMP location are not regular and the robot needs some time to reach the regular walking cycle. The CM crosses the region which is much larger than the allowed ZMP region. This means that walking with given step size, walking speed and pelvis height is possible only due to inertial forces. By realization of this gait pattern with a low speed, the robot will rotate around the foot edge.

To mention is that without using eqs. (5) for setting kinematic parameters it was not possible to achieve stable periodic movement with specified task parameters and given robot model.

## Non-periodical Motion

The same algorithm can be used for non-periodical motion like sitting down and standing up. Often the system is in rest after non-periodical motion. In this case, the kinematic parameters which define the end configuration of the system, should guarantee the statical stability of the system. For this the control of position of CM was used. This approach and the simulation were presented by the authors in [4].

## Model Parameters Uncertainty

The proposed algorithm can be used for both, as trajectory generator and as a controller of the system. In the second case, the algorithm runs in the loop with a real system. If the parameters of the real system do not match the parameters of the model, the behavior of the system which was linearized and decoupled by means of inverse dynamic, differs from the behavior of a double integrator. This disturbs the generation of the motion patterns or even makes the stable motion impossible.

To make the algorithm insensitive to parameters uncertainty a sliding mode controller (SMC), see e.g.[7], was used. Each of the blocks *ctrl* in Fig. 1 is replaced by the chain of blocks shown in Fig. 3. The acceleration $a$ computed by block *ctrl* is fed to the double integrator which produces the desired trajectory $(x,\upsilon)$. The sliding mode controller, block *SMC*, forces the system to move along $(x,\upsilon)$ by generating the new acceleration $a^*$.

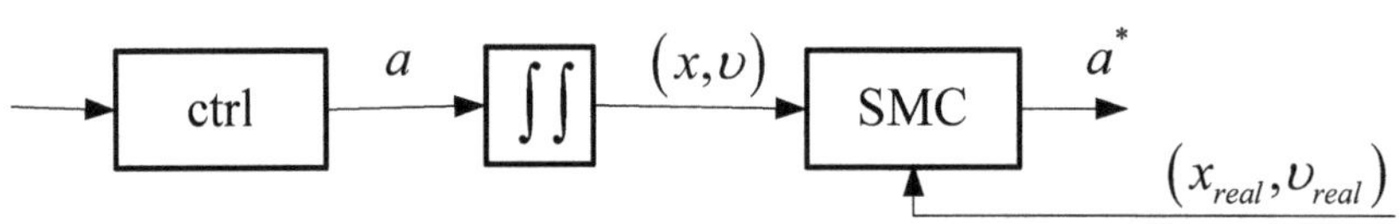


**Fig. 3.** Sliding mode controller enhances the parameter insensitivity.

The block SMC is implemented with following eq.:

$$a^* = -\int K \operatorname{sign}\left(C_x \Delta x + C_v \Delta \upsilon + C_a \Delta a\right) dt \tag{7}$$

The constants in eq. (7) for the presented simulation experiment were set as follows: $K = 500$, $C_x = 1$, $C_v = 0.66$, $C_a = 0.11$. For the step in the sign-function a linear approximation with the gain $g = 100$ was used.

More about SMC with eq. (7), choosing constant $K$ and computation of constants $C_x$, $C_v$, $C_a$ can be found in [8], [9].

## Conclusion

The proposed algorithm combines online generation of stable motion patterns and control of an exoskeleton. The main advantages of this algorithm are the minimal computational effort and possibility to change the motion parameters online during the movement. In the proposed algorithm the allowed range for ZMP was specified instead of its position or trajectory. The main disadvantage of the algorithm is its local nature. By very fast movement the projection operator is active for a relatively long period of time and the controllers in the feedback loop are not able to correct the disturbances in the motion pattern. In this case, the robot can be moved into a singular region. To deal with parameter uncertainties of the system in the case where the algorithm is used as a global controller, the application of the sliding mode controller was proposed. The simulation experiments show that for scenarios relevant in practice the algorithm is able to produce stable motion of a 3D model of a biped robot.

### References

[1] F. Pfeiffer, K. Löffler, M. Gienger, "The concept of jogging Johnie", in *Proc. of IEEE Int. Conf. on Robotics & Automation*, 2002.

[2] M. Vukobratovic, B. Borovac, D. Surdilovic, "Zero-moment point – proper interpretation and new applications", in *Proc. of the IEEE-RAS Int. Conf. on Humanoid Robots*, 2001.

[3] M. Spong, W. Vidyasagar, *Robot Dynamics and Control*. John Wiley & Sons, 1989.

[4] K. Kondak, G. Hommel, "Control and Online Computation of Stable Movement for Biped Robots". In *Proc. of IEEE/RSJ Int. Conf. on Intelligent Robots and Systems*, 2003.

[5] K. Kondak, G. Hommel, "Control algorithm for stable walking of biped robots". In *Proc. of the 6th Int. Conf. on Climbing and Walking Robots (CLAWAR)*, 2003.

[6] K. Kondak, G. Hommel, "Online Generation of Stable Gait for Biped Robots with Feedback Loop Algorithm". In *Proc. of the IEEE Conf. on Robotics, Automation and Mechatronics*, 2004

[7] V. Utkin, *Sliding Modes in Control Optimization*, Springer-Verlag, 1992.

[8] K. Kondak, G. Hommel, "Robust Motion Control for Robotic Systems Using Sliding Mode", proposed to the *IEEE/RSJ Int. Conf. on Intelligent Robots and Systems,* 2005.
[9] A. Wege, K. Kondak, G. Hommel, "Development and Control of a Hand Exoskeleton for Rehabilitation Purpases". In Proc. of Int. Workshop on Human Interaction with Machines, 2005.

# A Study on Situational Description in Domain of Mobile Phone Operation

Feng Gao, Yuquan Chen, Weilin Wu, Ruzhan Lu

Department of Computer Science and Engineering, Shanghai Jiao Tong University, Shanghai, China, 200030
gaofeng@cs.sjtu.edu.cn

**Abstract.** The natural language interpretation in Intensional Logical Model (ILM) consists of three steps: basic meaning description, concepts combination and situational description. This paper focuses on the general approach to and process of situational description, including the application of the ILM approach to the conceptual combination in the field of dialogue. In the paper, the mobile phone operation is an instantiation of the human-computer interaction. In the first section, the task for the situational description will be outlined. In the second section, the dynamic and static conceptual combination will be discussed in the framework of ILM. And the computing model will be introduced for situational description in third section.
**Keywords.** HCI, Intensional Logical Model, Situation.

## Introduction

The core of human-computer interaction is natural language understanding. But it is different from the general natural language processing in Information Retrieving, Information Filtering, and Automatic Summarization, because the utterances in the human-computer interaction are intentional, human and computer get information or direction in the process of interaction [John R. Searle 2001, Tu's preface, F25].Therefore, it is possible to adopt a functional approach to have an analysis of meaning for machine processing. That is to say, the meaning in the human-computer interaction is classified into two major types-information acquirement and command distribution, which is equivalent to the distinction between illocutionary act and perlocutionary act in ordinary language philosophy [John R. Searle 2001, Tu's preface, F28, F29, P54].



*G. Hommel and Sheng Huanye (eds.), Human Interaction with Machines,* 117-128.

Another motive for the application of functional approach is the uniqueness of Chinese language. Due to the lack of explicit morphological markers, the meaning compounding is completely driven by the meanings of the constituents. How to account for the compounding process is the bottleneck for the meaning computation in Chinese. Only combination order can be found from a structuralism approach. But this information of combination order is inefficient for the processing of meaning, especially intentional meaning. Typical questions such as multi-category and polysemy are difficult for the structuralism approach. [Lu 2001, P1]

Based on conceptual analysis, Intensional Logical Model(ILM) theory can solve the problem more efficiently [Song & Lu 2003, P252]. According to the theory, the linguistic units including words, sentences and discourse are concepts composed of their constituents through combinatorial computation such as mapping, inhabitation, or substitution in line with their intensional characteristics. Under the guidance of the theory, the processing of meaning in the domain of human-computer interaction is based on the logical framework of the intensional characteristics of the basic unit in Chinese such as characters or words. At the same time, a domain is restricted for situational description. So, the concept composed in the ILM model can be interpretable for the given task. [Mao & Gao 2003, P246]

## 1 The task of situational description

The purpose of situational description in Intensional Logic Model is to retrieve actual meaning from concept. So, it can not only be regarded as limiting the form of the utterance "what it can be", but decide the final meaning of the utterance. It is different from Situation Semantics(SS), where the final meaning of the utterance is represented as a True-False value. In SS, situation replaced True-False value as the goal of computation, but here, situation still has a spatio-temporal representation [Barwise Jon & John Perry, 1983]. Thus, the situational description in Intensional Logic Model must introduce concrete locations, persons and the restrictive depiction of the dialogue content. And at that time, time and space factors have already contained in the location, person and restrictions.

In the domain of human-computer interaction, the positions human (speaker) and computer (listener) are not the same as those in common dialog. Human is always putting forward the requirements, including executing functions (such as dial the phone number, send messages, set control panels, etc) and retrieving information (such as the content of messages, the phone number of someone, time, battery remained, etc), while computer is just accept passively and never actively hint the service or information from human until the command of user is ambiguous. Therefore the final semantics of the whole dialog can be represented as:

| 人 | 向 | 计算机 | 索取 | 服务 |
|---|---|---|---|---|
| Human | from | computer | acquire | service |

Where service can be represented as:

| 执行 | 功能 | 或者 | 返回 | 信息 |
|---|---|---|---|---|
| Executing | functions | or | return | information |

Information is related to some specific attributes and value. For example, human name, place name and institute name are the identifier of some number, and time, hour and minute are the identifier of the "current time", "on" and "off" are the settings of some controllers. Then, service can be further represented as:

| 执行 | 功能 | 或者 | 返回 | 属性 | 的 | 属性值 |
|---|---|---|---|---|---|---|
| Executing | functions | or | return | the attribute | 's | values |

When human give requirements to computer, besides some direct commands, such as ordering something directly ("give me …"), some indirect commands can also be made. For example, utterances like "I want …" or "I need …" represents the intentions of the speaker originally, but these intentions become commands because of the active-passive relationship between human and computer in the dialog. Thus, the final semantics of the whole dialog can be represented as:

| 人 | 表达 | 向 | 计算机 | 索取 | 服务 |
|---|---|---|---|---|---|
| Human | express the idea | from | computer | acquiring | service |

If the final semantics is regarded as a continuous process, the "human", "express", "computer", "execute", "function", "retrieve", "attribute" and "value" are the observation results of some angle of the whole process, and the whole process can be recognized through these description. Thus, the goal of semantic analysis is to retrieve the process (including the observation results above) from the utterance.

# 2 Intensional Logic analysis of concept

## 2.1 Intensional logical model for nouns

### *2.1.1 Logical meaning of nouns*

Polysemy refers to the phenomenon that different sememes of a word are interrelated yet different from each other. The meaning of a noun is one of its sememes in the actual realization. Therefore, there exists the difficulty of disambiguation for the computer. For example, *telephone* is a polysemy:

Ex.1) 张教授 的 电话
prof. Zhang 's telephone
(The speaker want the computer to provide the telephone number of Professor Zhang.)

Ex.2) 张 授 的 电话
prof. Zhang 's telephone
A call from Prof. Zhang

Ex.3) 张 授 的 电话
prof. Zhang 's telephone
A call for Prof. Zhang

Ex.4) 张 授 的 电话
prof. Zhang 's telephone
The telephone of Prof. Zhang (The original meaning of "telephone".)

Seen from the instance above, the word "telephone" has additional meanings of "telephone number" and "the communication process " besides its original meaning of "telephone machine" (a kind of communication device). The formation of a noun concept needs its hypernymous concept on which it depends, the special characters to distinguish itself from the concept, and every markers related to the character. So, the semantic framework of "telephone" can be described as the following:

*[telephone]: a kind of communication device that is used by users to transmit voice from one to another .*

In this framework, the markers of "user" and "another one" may be names of human, organization, group and address. The marker of "wire" can be number.

### 2.1.2 Semantic framework of nouns

According to the analysis above, the noun semantic framework ought to be described as $N =< H, F, C_1, C_2, ..., C_n >$, where H is the hypernymous concept of W, and F is the character description to distinguish itself from H, and $C_1, ..., C_n$ are relative markers.

Because the extension will be narrowed and intensional attributes will be increased when combining two nouns , the semantic composition process may be viewed as defining process of all uncertain $C_i$ of F.

Suppose that the semantic category of $C_i$ is $Cat(C_i)$, so we have the composition rules applied to $N_1.N_2$.

$$\textit{If } Cat(N_1) \in Cat(C_i(N_2)), \textit{ THEN } M(N_1.N_2) \equiv F(N_2)(N_1)$$

Where $F_N$ represents the intensional attribute of $F_N$, and $F_N(N_1)$ represents the application of N's intensional attribute into $N_1$.

In "Axiom of Identifical", Searle presents that there exit three ways to perform the denoting function of a word [John R. Searle 2001, Tu's preface, F30], which points to the objects denoted by the word in the real world. Accordingly, when language (including charaters, words and phrases) denotes concepts in the conceptual world, there are the following three ways:

1. The words spoken by a speaker must contain the correct predicate of object denoted by some concept.
2. When a speaker employs words to express some concept in certain context, it is guaranteed to provide notional markers in order to tell this concept from other concepts of the same words, which include objects, actions and index words.
3. With the performance of any function, his utterance performs also the expressive function of the concept.

As for the first way, it depends on verbs to determine the seme of noun combined with these verbs. As for the third measure, the complete meaning of the nominal structure which lacks in predicate will be provided in a given context and situation. The notion in the second way includes objects, actions and words. For the computer processing, only words can be used as resources for understanding, so it is certain to lead to ambiguity in understanding as expressed in examples above. Our mission is trying our best to provide every impossible semantic understandings as a premise for the further interaction to search for sememes.

Taking "telephone" as an instance, when the previous word is "Professor Zhang" with the concept of "human", it is needed to research the intensional attributes of "telephone" in order to determine which marker is to be used to modify "telephone". In its intensional attributes,

Both "user" and "the other side" can be the marker of human beings. Thus, "Professor Zhang" can not only modify the telephone (in example 2) as "users", but also modify "the telephone" as "the other side" (in example 3).

Besides the intension (identification) relation, there exists a pervasive extensional relation called 'ownership'. "Prof. Zhang" meets the requirement to be the holder in the relationship. He can own "telephone" (in example 4), as well as the "circuit" of the "telephone" which is marked by the number (indexes) (in example 1).

## 2.2 Logic models of verbs

The verb is generally treated as the predicate in logic. But there are many multi-category words crossing nouns and verbs in Chinese, such as "fax" and "BP". Besides these words, a VP can be a NP, such as "comes a call", and "writing back". Actually, the prepositions in Chinese evolved from verbs so that it is helpful to treat prepositions as verbs.

### *2.2.1 Sememes of mono-category verbs*

The sememes of the mono-category verbs can be treated as a predicate directly. Since we consider the conceptual combination, it is unnecessary to take into account the problem of intension and extension in truth value semantics. The new problem is how to deal with the two or more nouns in consequence. Taking dialing for example, the verb "dial" can be followed by the noun "telephone" as well as the name of a person, place, organization including address and a general name for the organizations. For example:

Ex.5) *拨打 电话*
*Dial telephone*
Dial

Ex.6) *拨打 张教授*
*Dial Professor Zhang*
Dial Professor Zhang

Ex.7) *拨打 张教授 的 电话*
*Dial Professor Zhang 's telephone*
Dial Professor Zhang

Ex.8) *拨打 张教授 办公室*
*Dial Professor Zhang office*
Dial the office of Professor Zhang

Does the order of combination have influence on the semantics of the whole VP structure? We hope not. In spite of the syntactic differences, the concepts(intensional features) remain the same no matter the two nouns combine first and then combine with the verb, or the verb combines with the previous noun to modify the following noun.

To realize the goal, information should be added in the description of the noun in the first way as exampled in "telephone". At the same time, the definition of the verbal structure should be enlarged so that the definition of the verb should not be confined into the description of an action but extended to describe the events of the action. We call them "call by body " and "call by name" respectively.

### 2.2.2 Meaning of V-N multi-category words

The semantic roles of V-N multi-category words were classified into three types including Agent, Result and Tool [Liu Shun 2003, P22]. The extended semantic description of the verb will be concerned with the information in these aspects. The following distinction is according to the semantic roles of the nouns:

Agent : Such as "editor", which is the person engaged in editing

Tool: Such as "locking" , which is a tool needed to perform the action of locking

Result: Such as "telephoning", and "writing back", which is "calling back" and "returning mail" respectively.

If the process of action described by the verb is viewed as a dynamic and changeable process, Agent, Tool and Result are the image of this process from the three different angels. This is in accordance with a philosophical distinction between SPAN/SNAP [Smith, B.,1997]. Thus, the problem of multi-category is converted into a problem resulted from the different perspectives and ways of the same process.

### 2.2.3 Sememe of V-P multi-category word

All prepositions in Chinese evolved from verbs. This is a result from abstraction and generalization of the object and result connected with the verb. So the prepositions are used like verbs basically. From a linguistic point of view, it is V-P multi-categoric. The basis for the generalization or abstraction is the abstract circumstances in which utterances happen. For example, the sememe of "give" as a verb is "to make others get or suffer". And the sememe of the prepositions is to introduce the object of the action or the purpose or agent of the behavior. If the indirect object (patient) of "give" as a verb extends from object (static) to dynamic process (dynamic), then it corresponds to the action and behavior of the preposition. For example:

Ex.9) 给 我 张教授 的 电话 号码

*Give me Professor Zhang 's telephone number*

Tell me the telephone number of Professor Zhang

Ex.10) 给 张教授 打 个 电话

*Give Professor Zhang call a telephone*

Please call Professor Zhang.

If we view the action process "give sb. a call" in Ex.10) the same thing as "telephone number", then "give" in 10) also means "make sb. get", thus the whole b) can be interpreted as "Professor Zhang get the treatment of a phone call" , which is consistent with interpreting 10) as a command.

# 3 Computing

## 3.1 Background of Language Theory

### 3.1.1 Definition of words

The composition in Chinese is achieved through the meaning of the components. Because of lacking morphological markers, as exemplified by the above examples, a word form inevitably has many senses. First, we need to guarantee that we describe the various senses of the word, and then we can eliminate the improper senses in the re-composition process step by step and finally get the final reasonable result.

There is a philosophical theory that things have two different views: SNAP view and SPAN view. Similarly, the senses of a Chinese word are the result of the common effect of the basic meaning (a process combined of the beginning, developing, and finishing phases of a thing, called P-view, similar as SPAN) and angle of the "Observation" process(N-view, .the observation result of a process profile, similar as SNAP).

### 3.1.2 Meaning of concept composition

Concept composition can have two kinds of relations, one is that the anterior concept modify the posterior concept which corresponds to determining a specific content when the posterior concept's observation angle is not determined; the other kind is that the posterior concept modifies the anterior concept, which differentiate with the first one in that it views this modification as an event and the intension of the event is "The posterior concept modifies the anterior concept".

"Modification" can be viewed as applying a modifier as a determined value to the intension of the concept modified. This determined process accords to the consistence between the category of the needed concept and the category of the concept being modified.

Because the variety of the intensional properties of Nouns, it is a dilemma as to how to determine the consistence between categories. On the one hand, if we use the various values of observation profile of a noun to match categories, a structure will inevitably produce multiple composition meanings, just as "telephone" in "Professor Zhang's Telephone"; on the other hand, if we only use the basic meaning of a noun as the matching category, the right senses can't be obtained, just as "office" in "Professor Zhang's office" whose hyponymy is "Location", but in specific domain, telephone number of the location is needed. It is based on the following rules:

1. The device is associated with its serial number. (The principle of naming)
2. The device is associated with its location. (The principle of entity-space/location association)
3. The serial number is associated with a location. (from 1, 2 and the rule of transition)
4. The intensional properties of the serial number are absorbed by the location or an observing angle of "场所/location" is added to the properties. (The heritage principle of intensional properties)

We have to note that the entity-space/location principle on which rule 2 depends, is conditional and therefore not general holding only in specific domains and can be viewed as domain knowledge. For example, the association between"办公室/office" establishes only in the domain, while the association between "办公室/office" and "地址/address" belongs to common knowledge: the principle of location naming.

### 3.1.3 The consistent processing of verbal complement's senses in combination

As mentioned in section 1. In a specific domain such as HCI, "我/I" can be identified as the speaker, so the sense of "给我/give me or for me", no matter whether we view "给" as a verb or as a preposition it always means "let the speaker get". Similarly, "你/you" can be identified as the "listener", which is the computer here. So all declarative utterances with "你/you" as the subject can be analyzed as command utterances with the speaker(user) as commander and the computer as the command receiver.

"想/want to", "要/would like to " can be manipulated in the similar ways.

The direct meaning of "我想/I want to…", "我要/I'd like to…" is to express the will of the speaker, while their actual pragmatic effect is to issue command to the listener, to make the listener do what the speaker wants. That the utterances have such functions is because that the inequality in status of speaker and listener in a specific HCI dialog process.

## 3.2 Definitions and Rules

### *Def 1. Definition of Concept in IML*

The extension of a concept is the characters which used to record it. The intension of the concept is the meaning assigned to these characters by psychological factors, social factors and historical factors, etc. The extension of a concept is unique. (It is just like denotation. Extension denotes a unique entity in real world.)

### *Def 2. P-view and N-view of a concept*

For any concept C, it's P-view (named as P(c)) is the set of all of its intensional properties while its N-view(named as N(c)) is the value got by viewing C's P-view from an observing angle.

In fact, it is an extension of using time and space as an observing angle in possible world semantics. Correspondently, their relation can be expressed as:

$N(c)=P(c)(i)$, where i  I, I is the set of all possible observing angles (referred to as possible mode).

***Def 3 P-view Transformation(P-Trans)***

For a N-view of any unit of utterance, P-view Transformation gets a P-view related with that N-view:

$P(C) = \{F| \exists C\ F(C)|_K = N(C)\}$, where $K \in I$, I is the set of Possible mode, I={production={reason, origin, source}, develop={manner, tool, procedure}, finish={result, purpose}。

***Def 4 N-view Transformation(N-Trans)***

For a P-view of any unit of utterance, N-view Transformation gets a N-view related with that P-view:

$N(C) = \{I|$ for any $K, K' \in I$, $\text{P-Trans}(P(C)|_K)|_{K'} = \text{P-Trans}(P(C)|_{K'})|_K\}$

***Rule 5 Quiz-rule***

If there are two or more P-view produced by P-Trans from a N-view, the question for explicit information will be created.

***Rule 6 Context-rule***

The N-view of the utterance will be filled into the P-view which has produced by former utterance.. The initial P-view of the utterance will be produce use the P-Trans from the N-view of the utterance.

***Rule 7 Satisfaction-rule***

The interaction will end if only if one and only one P-view of the situation has the explicit value in its possible modes.

## 4 Conclusion

The ideas of situational description in IML are discussed concisely in the paper. For a task of understanding (more accurately, interpreting) the utterance in HCI, the P-view of the domain should be given and the polysemy of the basic word in this domain should be described as the part of the situation. The aim of interpretation of whole dialogue is acquiring the value of possible modes of the P-view. The future work will focus on (1)the ordinary phenomena such as repetition, redundancy, rectification, etc which puzzle the current system in step of P-Trans and Satisfaction [Guo & Gao 2003, P575]; (2) the more effective approach to describe the sememes of polysemy.

## References


1. John R. Searle, 2001, *Speech Acts: An Essay in the Philosophy of Language.* Reprint edition published by Foreign Language Teaching and Research Press with the permission of the Syndicate of the Press of the University of Cambridge, Cambridge, United Kingdom.
2. Lu Ruzhan, 2001, *The The Interpretation of "SHI" in Intensional Logical Model,* The proceeding of JSCL-2001, The Press of Tsinghua University, Beijing, China.
3. Mao Jia-ju, Gao Feng, Chen Qiu-lin and Lu Ruzhan, 2003, *A Situation-theoretic analysis of VA-statements,* The proceeding of JSCL-2003, The Press of Tsinghua University, Beijing, China.
4. Barwise Jon, John Perry, 1983, *Stuation and Attitudes,* Cambridge, MA:MIT Press
5. Song Chunyang, Lu Ruzhan and Fang Xianghong, 2003, *Research on the Interpretation Pattern of Compound-Word Meaning by Generalized Meaning Abstract,* The proceeding of JSCL- 2003, The Press of Tsinghua University, Beijing, China.
6. Liu Shun, 2003, The Research on Multi-view of Modern Chinese Nouns, 学林出版社 Xue Lin Press, Shanghai, China.
7. Smith, B,1997, On Substances, Accidents and Universals: In Defence of a Constituent Ontology. Philosophical Papers, 26, 105-127.
8. Guo Rong, Gao Feng, Mao Jia-ju and Lu Ruzhan, 2003, Negation Processing in Human-machine Spoken Dialogue System, The proceeding of JSCL-2003,The Press of Tsinghua University, Beijing, China.

# Computer-Supported Decision Making with Object Dependent Costs for Misclassifications

Fritz Wysotzki and Peter Geibel

Methods of Artificial Intelligence, Sekr. FR 5-8, Faculty IV, TU Berlin, Franklinstr. 28/29, D-10587 Berlin, Germany

**Summary.** It is described, how object dependent costs can be used in learning decision trees for cost optimal instead of error minimal class decisions. This is demonstrated using decision theory and the algorithm CAL5, which automatically converts real-valued attributes into discrete-valued ones by constructing intervals. Then, a cost dependent information measure is defined for selection of the (locally) next maximally discriminating attribute in building the tree. Experiments with two artificial data sets and one application example show the feasibility of this approach and that it is more adequate than a method using cost matrices if cost dependent training objects are available.

**Keywords.** decision trees, learning, misclassifications, costs, object dependence

## 1 Introduction

The inductive construction of classifiers from training sets is one of the most common research areas in machine learning and therefore in man-computer interaction. The classical task is to find a hypothesis (a classifier) that minimizes the mean classification error [4]. As it is stated in the Technological Roadmap of the MLnetII project (Network of Excellence in Machine Learning [5]), the consideration of costs in learning and classification is one of the most relevant fields in future Machine Learning research. One way to incorporate costs is the usage of a cost matrix, that specifies the mean misclassification costs in a class dependent manner [4, 3, 1]. In general, the cost matrix has to be provided by an expert. Using the cost matrix implies that the misclassification costs are assumed to be the same for each example of the respective class. A more natural approach is to let the cost depend on the single training example and not only on its class.

One application example is the classification of credit applicants to a bank as either being a "good customer" (who will pay back the credit) or a "bad customer" (who will not pay back parts of the credit loan).



*G. Hommel and Sheng Huanye (eds.), Human Interaction with Machines,* 129-139.

A classifier for this task can be constructed from the bank's records of past credit applications that contain personal information on the customer, information on the credit itself (amount, duration etc.), on the back payment of the loan, and information on the actual gain or loss of the bank. The gain or loss in a single case forms the misclassification cost for that example in a natural way. For a good customer, the cost is the loss, if the customer has been rejected. For a bad customer, the cost is simply the actual loss that occurred.

Our claim is that using the individual costs directly is more natural than constructing a cost matrix and may produce more accurate classifiers. An approach for using example dependent costs has been discussed in the context of Rough Classifiers [2], the Perceptron, Piecewise Linear Classifiers and the Support Vector Machines [8, 10] and will be applied to the learning of decision trees in this paper (see also [9]). Decision trees (see for example [4]) have the advantage that they can be decomposed into a set of rules which can be interpreted in a humanlike fashion. Therefore decision tree learning can be used as a tool for automatic knowledge acquisition in man-machine interaction, e.g. for expert systems.

This article is structured as follows. In section 2, the algorithm CAL5 for induction of decision trees is introduced. Section 3 describes the modification of CAL5 in case of object dependent costs and section 4 a cost dependent information measure for selecting the attributes for tree building in a (locally) optimal manner. Experiments on two artificial domains and on a credit data set can be found in section 5. Section 6 concludes the article.

## 2 Decision Tree Learning by CAL5

The algorithm CAL5 [4, 6, 7] for learning decision trees for classification and prediction converts real-valued attributes (features) into discrete-valued ones by definition of intervals on the real dimension using statistical considerations. The intervals (corresponding to discrete or "linguistic" values) are automatically constructed and adapted to establish an optimal discrimination of the classes by axis-parallel hyperplanes in the feature space. The trees are built top-down in the usual manner by stepwise branching with new attributes to improve the discrimination. An interval on a new dimension (corresponding to the next test refining the tree on a given path) is formed if the hypothesis "one class dominates in the interval" or the alternative hypothesis "no class dominates" can be decided on a user defined confidence level by means of the estimated conditional class probabilities. "Dominates" means that the class probability exceeds some threshold given by the user.

In more detail: If during tree growing a terminal node representing an interval in which no class dominates is reached, it will be refined using the next attribute $x$ for branching. It is selected using the transinformation measure (sect. 3) to decide which is the (locally) best discriminating attribute, i.e. which transfers maximal information on the classes. In the first step of the interval construction all values of the training objects reaching the terminal node are ordered along the new dimension $x$. In the second step values are collected one by one from left to right forming intervals tentatively. Now a confidence interval $[\delta_1, \delta_2]$ for each class probability is defined for a user given confidence level $1-\alpha$ for estimating the class probabilities in each current tentative interval $I$ on $x$. It depends on the relative frequency $n_k/n$ of class $k$, the total number $n$ of objects in $I$ and on $\alpha$. A Bernoulli distribution is assumed for the occurrence of class symbols in $I$. Then the following "metadecision" is made for each tentative $x$-interval $I$:

1) If for a class $k$ in $I$

$$\delta_1(n_k/n, n, \alpha) \geq S$$

holds, then decide "class $k$ dominates in $I$".
$S \leq 1.0$ is a user-given threshold, defining the maximal admissible error $1-S$ in class decision using the tree. If class $k$ dominates in $I$, the path will be closed and $k$ is attached as a class label to the newly created terminal node.

2) If for all classes $k$ in $I$

$$\delta_2(n_k/n, n, \alpha) < S,$$

i.e. no class dominates, the tree is refined using the next attribute.

3) If neither 1) nor 2) holds, a class decision is rejected, too, and $I$ is extended by the next value in the ordering on $x$ to enlarge the statistics. The procedure is repeated recursively until all intervals (i.e. discrete values) of $x$ are constructed.

The confidence interval $[\delta_1, \delta_2]$ is defined using the Chebychev-Inequality. A special heuristics is applied in the case that (due to the finite training set) there remain intervals on dimension $x$ fulfilling neither hypothesis 1), nor hypothesis 2) (see sect. 3). Finally, adjacent intervals with the same class label are joined. If there are a priori discrete valued attributes the metadecision procedure described above is applied simply to the set of objects reaching the terminal node in question.

If costs for not recognizing single classes are given, CAL5 uses class dependent thresholds $S_k$. From decision theory it follows [6] that one has to choose

$$S_k \approx const/cost_k,$$

where $cost_k$ is the cost for misclassification of an object belonging to class *k*. The cost must be provided by the user and depends on the class only. The main aim of this paper is to introduce object depending costs given with the objects of the training set. This allows to use them locally for an interval *I* (defining a region in the feature space) and independently from subjective estimates of experts. This is explained in more detail in the next sections.

## 3 Using Cost Dependent Decision Tresholds in Case of Object Dependent Costs

Now we extend the metadecision rules 1)-3) introduced in sect. 2 for the case that the training objects (described by feature vectors) are presented with costs for misclassification, i.e. we have a training set

$$(x^{(p)}, c^{(p)}, k^{(p)}), p = 1, \ldots, r$$

with feature (attribute-) vector $x^{(p)}$, class $k^{(p)} \in \{kj, j = 1, \ldots, n\}$ and costs of misclassification $c^{(p)}$.

We consider case 2) of the metadecision rules of sect. 2, where a new branch at some final node of the decision tree has to be constructed with attribute *x* selected. Assume that *I* is the (perhaps temporary) current interval considered representing a *discrete* value of *x*. In what follows we write *x* for both the original real-valued attribute *x* and the discrete-valued one to be constructed. *x* and class variable *k* can also be interpreted as stochastic variables in the context of probability theory. From decision theory the following rule for decision for a class can be derived [6]:

Decide class $k_j$ if

1) $c_j(I)p(k_j/I) > c_i(I)p(k_i/I)\ \forall i \neq j$ (Bayes rule)

and simultaneously

2) $p(k_j/I) > (\sum_{i=1}^{m} c_i(I)p(k_i/I) - c_0(I))/c_j(I) =_{Def.} S_j(I).$

$p(k_j/I)$ is the conditional probability of $k_j$ (to be approximated by the relative frequency) in interval *I*. Note that *I* depends on the path from the root of the current decision tree to the root node of the new branch (labelled with

$x$) where it was constructed. $c_j(I)$ is the mean cost for misclassification of class $k_j$ in interval $I$, i.e.

$$c_j(I) = \sum_{\substack{x(p) \in I \\ x(p) \in kj}} c(x^{(p)}) / n_j(I)$$

where $n_j(I)$ is the number of training vectors $x^{(p)}$ of class $k_j$ falling into $I$. $S_j(I)$ is the decision threshold, now dependent on the mean cost $c_j(I)$ in $I$. $c_0(I)$ is the cost for the rejection of a dominance decision and therefore further branching. Since it is not known in advance we eliminate the sum containing it and get for all $i, j$

$$S_j(I)/S_i(I) = c_i(I)/c_j(I)$$

and therefore

$$S_j(I) = c_{min}(I) S_{max}(I)/c_j(I)$$

$c_{min}(I)$ is the cost of the class with minimal cost in $I$, and $S_{max}(I)$ the corresponding threshold, which must be defined by the user. With the $S_k(I)$ we have the modified metadecision rules (see sect. 2):

1) If for a class $k_j$ in $I$

$$\delta_1(n_j/n, n, \alpha) \geq S_j(I)$$

holds, then decide "$k$ dominates in $I$".

2) If for all classes $k$ in $I$

$$\delta_2(n_j/n, n, \alpha) < S_j(I)$$

i.e. no class dominates, a branch with the next attribute is constructed. In case 3) the description given in sect. 2 holds without modification.

## 4 Computation of a Cost Dependent Transinformation Measure

In case 2) the next attribute for branching has to be selected. CAL5 uses the attribute $x$ giving the (local) maximal information after discretization, i.e. the maximum reduction of uncertainty on the classes. This so called transinformation measure $I_x$ is defined as the difference of the entropy $H_k$ of the classes before measuring $x$ and $H_{k/x}$, the expectation (mean) of entropy on the set of measured features (values) of $x$. We consider in this context $x$ and $k$ as stochastic variables with the values $x_i, i=1,...,m$ and $k_j$, $j=1,...,n$, respectively. Also, we write $c_{k_j}$ for the cost of misclassification of class $k_j$ instead of $c_j$ used in sect. 3. Then the transinformation is defined by

$$I_x = H_k - H_{k/x} = -\sum_{j=1}^{n} p(k_j) \log p(k_j) + \sum_{i=1}^{m} p(x_i) \sum_{j=1}^{n} p(k_j/x_i) \log p(k_j/x_i).$$

To define a cost dependent transinformation measure we introduce cost dependent probabilities $p^c(k_j)$, $p^c(k_j/x_i)$, $p^c(x_i)$ [8, 9] to replace $p(k_j)$, $p(k_j/x_i)$, $p(x_i)$in $I_x$ by the definitions:

$$p^c(k_j) =_{Def.} Bk_j p(k_j) / \sum_{j=1}^{n} Bk_j p(k_j) = Bk_j p(k_j)/b$$

with $B_{k_j} =_{Def.} \sum_{i=1}^{m} c_{k_j}(x_i) p(x_i/k_j)$ and $b =_{Def.} \sum_{j=1}^{n} B_{k_j} p(k_j)$

and $p^c(k_j/x_i) =_{Def.} c_{k_j}(x_i) p(k_j/x_i)/B_{x_i}$

with $B_{x_i} =_{Def.} \sum_{j=1}^{n} ckj(xi)p(kj/xi)$

and $p_c(x_i) =_{Def.} B_{x_i} p(x_i) / \sum_{i=1}^{m} B_{x_i} p(x_i) = B_{x_i} p(x_i)/b$

(It can be proved that $b = \sum_{j=1}^{n} Bk_j p(k_j) = \sum_{i=1}^{m} Bx_i p(x_i)$.) Note that the new probabilities modified by costs are defined by multiplying the original ones by the normalized costs.

## 5 Experiments

### 5.1 Experiments without Costs

In fig. 1 a training set (NSEP) is shown which consists of two overlapping classes, taken from a Gaussian distribution in the $(x_1,x_2)$-plane [9]. The attribute $x_1$ is irrelevant, i.e. the classification does not depend on it and can be performed with $x_2$ alone. The original CAL5-algorithm with parameters $\alpha=0.3$and $S_{max}=0.6$(and no costs) constructs two discriminating straight lines parallel to the x-axis. The region between them indicates that due to nonseparability no statistical safe decision is possible in it and a "majority decision" must be made.

The data set (MULTI) shown in fig. 2 consists of two classes containing two subclasses (clusters) each. All four clusters are taken from Gaussian distributions. CAL5 constructs two piecewise linear functions discriminating the classes (subdivisions within the same class built by CAL5 are omitted).

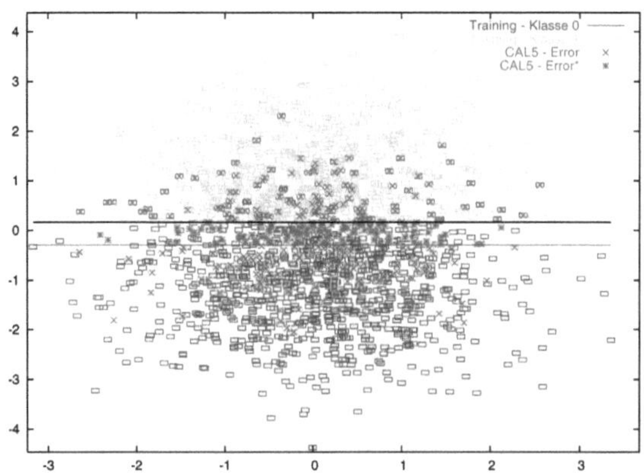


**Fig. 1.** Classifer for data set 'NSEP' (without costs).

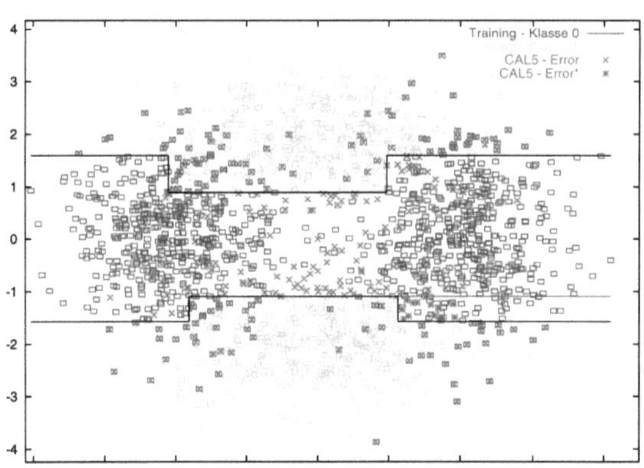


**Fig. 2.** Classifier for dat set 'MULTI' (without costs).

## 5.2 Experiments with Object Dependent Costs

From the data set NSEP (sect. 5.1) a new data set NSEP_OC is constructed using objects dependent costs. They are computed from the cost functions

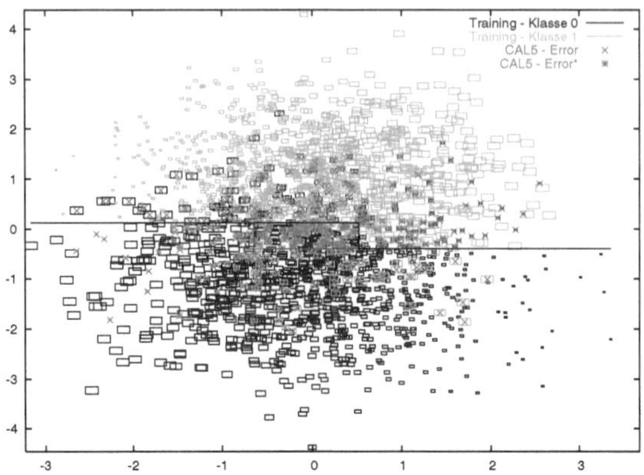


**Fig. 3.** Classifier for dat set ‘NSEP_OC’ (with object dependent costs).

$$c_0\,(x_1\,,x_2\,)=2/(1+exp(x_1\,))\text{ for objects of class 0}$$
$$\text{and } c_1\,(x_1\,,x_2\,)=2/(1+exp(-x_1\,))\text{ for objects of class 1}$$

both dependent on the irrelevant feature $x_1$ only.

In fig. 3 one can see the training data set and the class discrimination function computed by using the modified CAL5 (CAL5_OC) which includes cost functions “learned” from the object dependent costs given in the training set as described in sect. 3. The value of the appropriate cost function for a training object is indicated by the size of the rectangles representing the training objects.

Now the discrimination function is piecewise linear: the decision region for class 1 is enlarged in the positive halfspace of attribute $x_1$ and for class 0 in the negative halfspace of $x_1$, respectively, i.e. for high costs of misclassifications. That means the originally irrelevant attribute $x_1$ becomes now relevant for class discrimination because of the cost dependence, i.e. the decision tree is optimized not with respect to minimal classification error (as before) but with respect to minimal cost for misclassification. The costs defining the (local) decision thresholds are “learned” from the individual costs given with the training objects.

For the sake of comparison we an CAL5 without costs and CAL5_OC with object dependent costs with the (optimized) parameters $\alpha$ =0.3, S =0.6 in the first and $\alpha$ =0.2, $S_{max}$= 0.6 in the second case. As a result we got a mean cost of 0.165 in the first and 0.150 in the second case. This means a lowering of the mean cost by about 9% in the object dependent, i.e. optimized case is achieved. The classification error increases from 16.45% to16.85% [9]. Note that the classification error must increase for classes with low costs of misclassification.

For the data set MULTI the cost functions

$$c_1(x_1,x_2) = 1 \text{ for class } 1$$
$$\text{and } c_0(x_1,x_2) = 2/(1+exp(-x_1)) \text{ for class } 0$$

are chosen. Fig. 4 shows the resulting data set and the discrimination functions defining the decision regions for both classes.

Comparing this with fig. 2 (cost independent case) one can see the enlargement of the decision region containing the right cluster of class 0 consisting of objects with high costs for misclassification.

## 5.3 Application in a Real World Domain: German Credit Data Set

We conducted experiments on the German Credit Data Set [4, 8] from the STATLOG project. The data set has 700 examples of class "good customer" (class +1) and 300 examples of class "bad customer" (class -1). Each example is described by 24 attributes. Because the data set has no example dependent costs, we assumed the following cost model: If a good customer is incorrectly classified as a bad customer, we assumed the cost of 0.1**duration***amount*/12, where *duration* is the duration of the credit in months, and amount is the credit amount. We assumed an effective yearly interest rate of 0.1, i.e. 10% for every credit, because the actual interest rates are not given in the data set. If a bad customer is incorrectly classified as a good customer, we assumed that 75% of the whole credit amount is lost (normally, a customer will pay back at least part of the money). In the following, we will consider these costs as the real costs of the single cases. In our experiments, we wanted to compare the results using example dependent costs with the results, when a cost matrix is used. For the German Credit Data Set human experts estimated a relative cost of 5 for not recognizing a bad customer (class -1) as to be compared with a cost of 1 for not recognizing a good customer (class +1) and cost 0 in the case of correct classification (cost matrix). Since the costs in this cost matrix cannot directly be compared with the individual costs, we constructed a second cost matrix by computing the average of the costs of class +1 and class -1 examples of the training set, getting 6.27 and 29.51, respectively (the credit amounts were normalized to lie in the interval [0,100]). The ratio of the classification costs in the new cost matrix corresponds approximately to the ratio in the original matrix. We run CAL5 with the cost German Credit Data Set with the (optimized) parameters $\alpha = 0.25$ and $S_{max} = 0.93$. As a result we got a mean cost of 3.34 in the first and 2.97 in the second case. This means a lowering of mean cost by about 12% for the object dependent, i.e. optimized case. The classification error increases from 38.2% to 41.5% [9].

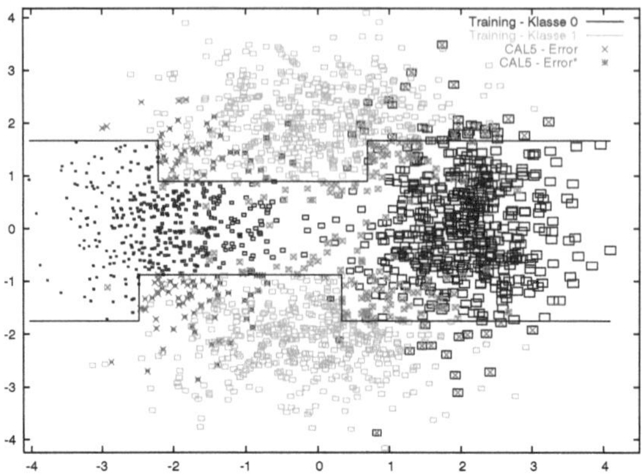


**Fig. 4.** ClassiÞer for dat set ÕMULTI OCÕ (with object dependent costs).

# 6 Conclusions

In this article we described, how object dependent costs can be used to learn decision trees for cost optimal instead of error minimal class decisions. This was demonstrated using decision theory and the algorithm CAL5, which automatically converts real-valued attributes into discrete-valued ones by constructing intervals. A cost dependent information measure has been defined for selection of the (locally) best next attribute in building the tree. Experiments with two artificial data sets and one application example show the feasibility of our approach and that it is more adequate than a method using cost matrices, if cost dependent training objects are available. Comparison with an extended perceptron algorithm and piecewise linear classifier, modified to include object dependent costs, too, [8] and applied to the same data sets used in our experiments described in sect. 5, shows consistent results, i.e. similar decision areas. The only difference is that decision trees separate the classes in the feature space by axis-parallel hyperplanes, whereas the abovementioned perceptronlike-algorithms use hyperplanes in general position, i.e. may construct better approximations of the class regions.

# References

1. C. Elkan. The foundations of Cost-Sensitive learning. In Bernhard Nebel, editor, *Proceedings of the seventeenth International Conference on Artificial Intelligence (IJCAI-01)*, pages 973-978, San Francisco, CA, August 4-10 2001. Morgan Kaufmann Publishers, Inc.

2. A. Lenarcik and Z. Piasta. Rough classifiers sensitive to costs varying from object to object. In Lech Polkowski and Andrzej Skowron, editors, *Proceedings of the 1st International Conference on Rough Sets and Current Trends in Computing (RSCTC-98)*, volume 1424 of *LNAI*, pages 222-230, Berlin, June 22 26 1998. Springer.
3. D. D. Margineantu and Th. G. Dietterich. Bootstrap methods for the cost-sensitive evaluation of classifiers. In *Proc. 17th International Conf. on Machine Learning*, pages 583-590. Morgan Kaufmann, San Francisco, CA, 2000.
4. D. Michie, D. H. Spiegelhalter, and C. C. Taylor. *Machine Learning, and Statistical Classification.* Series in Artificial Intelligence. Ellis Horwood, 1994.
5. Lorenza Saitta, editor. *Machine Learning – A Technological Roadmap*. University of Amsterdam, 2000. ISBN: 90-5470-096-3.
6. S. Unger and F. Wysotzki. Lernfähige Klassifizierungssysteme (Classifier Systems that are able to Learn). Akademie-Verlag, Berlin, 1981.
7. W. Müller and F. Wysotzki. The Decision-Tree Algorithm CAL5 Based on a Statistical Approach to Its Splitting Algorithm. In: G.Nakhaeizadeh and C.C.Taylor (Eds.), *Machine Learning and Statistics*, 45-65, Wiley, 1997.
8. P. Geibel and F. Wysotzki. Learning Perceptrons and Piecewise Linear Classifiers Sensitive to Example Dependent Costs. Applied Intelligence 21, 45-56, 2004.
9. A. Bendadi, O. Benn, P. Geibel, M. Hudik, T. Knebel and F. Wysotzki Lernen von Entscheidungäumen bei objektabhäbhängigen Fehlklassifikationskosten. Technischer Bericht Nr. 2004-18, TU Berlin, 2005. ISSN: 1436-9915.
10. P. Geibel, U. Brefeld and F. Wysotzki Perceptron and SVM Learning with Generalized Cost Models. Intelligent Data Analysis 8/5, 439-455, 2004.

# Robust Analysis and Interpretation of Spoken Chinese Queries

Ruzhan Lu, Weilin Wu, Feng Gao, Yuquan Chen

Department of Computer Science and Engineering, Shanghai Jiaotong University, Shanghai, China, 200030

**Abstract.** In spoken dialogue systems, a robust treatment is required for the input utterance due to the recognition errors and a large set of spontaneous speech phenomena. This paper describes a framework for robust analysis and interpretation of spoken language. This framework includes inherently robust mechanisms, i.e., robust semantic parsing and robust interpretation, and an additional component as well, i.e., recognition error correction. A preliminary evaluation demonstrates that a good robustness is achieved efficiently under this framework.
**Keywords.** Robust Analysis, Error Correction, Spoken Dialogue System

## 1 Introduction

In the information inquiry system via voice, the input utterances are mostly grammatically incorrect or ill-formed. Besides recognition errors (word insertion, omissions and substitution), spoken language is also plagued with a large set of spontaneous speech phenomena as follows [1, 2]: (1) Out-of-vocabulary words; (2) User noise - heavy breath noise, cough, filled pause, lips smacks, etc; (3) False start and self correction; (4) Repetitions, hesitations and stammering; (5) Extra-grammatical constructions - ellipse, out-of-order structure, etc. Most of these phenomena are to be handled by the linguistic processing component except that the first one is a problem of the speech recognizer. In essence, the approaches to address these phenomena can be divided into two classes [3]:

- **Inherently robust mechanisms**
  Embedded with fuzzy strategies, this approach handles exceptional inputs in the same way as regular inputs. It is more uniformly robust and systematic, but may involve the system becoming less responsive.
- **Requiring additional components**
  Given exceptional inputs, the processing steps are as follows: identify problem phenomena, analyze them and formulate specific responses.



*G. Hommel and Sheng Huanye (eds.), Human Interaction with Machines,* 141-147.

This approach is highly phenomenon specific: a satisfactory result can be achieved if the problem phenomenon is correctly identified, however, it also takes a risk that the identification process will misfire and employ strategies that are not required.

This paper describes a framework for robust analysis and interpretation of spoken language, which combines the two kinds of approaches stated above. In this framework, the components related to the robust processing include recognition error correction, which is an additional component, robust semantic parsing and robust interpretation, which are inherently robust mechanisms.

## 2 The System Architecture

The system architecture is shown in Figure 1. The diagrams in the shaded background represent the robust processing components. The recognized Chinese text is firstly passed to the preprocessor (segmentation and semantic classes tagging). Then, the robust semantic parser, accepts the results generated after some recognition errors are corrected, and produces the parse. The parsing result is then repaired again if possible. Finally, the interpretation component maps the repaired parse into a semantic representation. Here, the error correction breaks into two subcomponents: non-word error correction based on edit distance and rule-based real-word error correction. Thus, two types of approaches to achieve robustness described above are combined into a uniform framework. Firstly, recognition error correction is integrated as additional components. Secondly, inside the semantic parser, partial parsing and word skipping are employed. Finally, the interpretation component makes extensive use of high level knowledge to robustly interpret the parse. The latter two mechanisms are inherently robust.

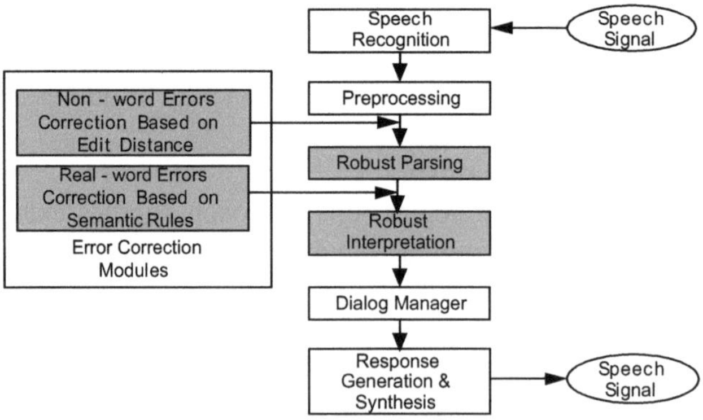


**Fig. 1.** The system architecture.

# 3 Recognition Error Correction

## 3.1 Types of Recognition Errors

These recognition errors can be categorized as non-word and real-word errors based on whether the corrupt string is a word or not. This kind of classification contributes a lot in choosing the strategies for the error detection and correction. Here, a recognition error is regarded as a non-word error if the corresponding string is not in the domain-specific dictionary, otherwise a real-word error.

## 3.2 Recognition Error Detection and Correction

### *3.2.1 Non-word Error Detection and Correction*

It is straightforward to detect the non-word errors. We can directly treat all the words with the tag UNKNOWN (this semantic tag stands for the out-of-vocabulary words) as the non-word errors. Continuous non-word errors are clustered into a single error.

The correction strategy for the non-word errors is based on the idea of edit distance. The procedure of error correction based on the edit distance includes: (1) to detect the errors; (2) to select the most similar strings for the corrupt string from the entries in the domain-specific dictionary.

The key point is to appropriately define the distance measure between the corrupt string and the corresponding correct string. For English strings, a practical metric is the MED (minimum edit distance). However, it is not suitable for Chinese strings since it only considers the similarity of the graphemes of two Chinese strings but can't reflect how similar their pronunciations are. Therefore, it is necessary to extend the notion of MED for Chinese strings. Assume a Chinese string $X_1^m$ can be transformed to another Chinese string $Y_1^n$ through $T_i$ insertions, $T_d$ deletions and $T_t$ transpositions, and the costs of the three operations are equal, we can define the corresponding distance as follows:

$$D(T_i, T_d, T_t) = (T_i + T_d + \sum_{j=1}^{T_t} C_t(j))/n \qquad (1)$$

In Equation (1), $C_t(j)$ is the cost of the $j^{\text{th}}$ transposition from a Chinese character in $X_1^m$ to the corresponding one in $Y_1^n$. Assume the pinyins of $X_j$ and $Y_j$ are $PX_j$ and $PY_j$ respectively, we can compute $C_t(j)$ as follows: $C_t(j) = MED(PX_j, PY_j) / Len(PX_j)$, where $Len(PX_j)$ is the length of $PX_j$. Finally, we can define the MED for Chinese strings as follows:

$$MED(X_1^m, Y_1^n) = \underset{T_i, T_d, T_t}{Min} (D(T_i, T_d, T_t)) \tag{2}$$

### 3.2.2 Real-word Error Detection and Correction

Based on the literature of recognition error correction [4] and the examination on the data we collected we know that real-word recognition errors occur in regular patterns rather than at random. Therefore, the rule-based approach is applied to deal with the real-word errors.

Here, the real-word errors correction mainly resorts to the semantic error correction rules. These semantic rules capture the semantic dependency among words. In example 1, the real-word error 从→乘 is caused by a character substitution. This error can be corrected by applying a semantic rule as follows:

^(semcls.$by, lex.乘)+(semcls.$loc) => (^.change.lex.从,
^.change.semcls.$from).

This rule is explained as follows: 乘 is corrected to 从 and change its semantic class to $from if the current word is 乘 and the next word is of the semantic class $loc.

| |
|---|
| 请问**乘**人民广场到外滩步行怎么走? * |
| 请问**从**人民广场到外滩步行怎么走? |
| How can I walk from the people's square to the Bund? |

**Example 1.** An example sentence with lexical semantic error.

## 4 Robust Parsing

The parser is the key component of linguistic processing. To account for the spontaneous input, the common practice is to extend the traditional parser to introduce robust behaviors. Modification of the parser can be divided into modification to the linguistic description that the parser applies, i.e., the grammar formalism, and modification to the parsing process, i.e., the parsing algorithm [3].

In our system, there is little modification for the linguistic description. The linguistic description fed into the parser is a context-free semantic grammar. In semantic grammars, non-terminals relate to semantic categories (e.g., [location], which are classified in terms of their function or semantic concept).

Conversely, there is wider room left for the modification to the main analysis process. The robust semantic parser in our system is extended from the chart parser. The first extension is that the condition for a complete analysis is relaxed, i.e., partial parsing serves as the robust parsing strategy. A chart parser is chosen here since all the previous partial parses are held as edges in the chart [1]. In fact, the input sentence of a spoken dialogue system often consists of well-formed phrases, and is often semantically well-formed, but is only syntactically incorrect [2]. Given the spontaneous input, the traditional parser often fails to construct a complete parse. However, these useful phrases in the spontaneous input, which capture the concepts intended by the user, can still be recognized through semantic-driven partial parsing. The other extension is to allow word skipping. Currently, the skipping words allowed include filled pauses (e.g. 嗯, 啊), out-of-vocabulary words and function words.

## 5 Robust Interpretation

Finally, the interpretation process translates a repaired parsing result into a semantic representation. Here the semantic representation is formalized as a frame with an internal structure consisting of slot/value pairs. Then, the task of interpretation can be viewed as slot filling using the task relevant information extracted from the parse tree.

There are various kinds of contextual information that can be clues for interpretation. Even if one of these clues misfire, a reliable interpretation can be still yielded according to the other clues. Therefore, we employ the strategy of combination of multiple clues described in [4] for robust interpretation. The first clue is the current utterance context. For instance, given the input sentence "我在外滩 - 去徐家汇 (I am in the Band - and go to Xujiahui)", we can conclude that "外滩 (the Band)" can be filled into a slot 'origin' because "去(go to)" signals a destination "徐家汇(Xuj iahui)". Another clue is the prior utterance from the system. If the system's prior utterance prompts the user to provide a 'destination', the location name in the user's reply utterance, which is not modified by any indicator word like "去(go to)", may be a destination. Finally, the dialog context is also useful information for interpretation. If the origin slot is already filled, the location name in the current utterance may be a destination when there is no other evidence. In the order from most significant to the least significant, these high level knowledge are ranked as follows: utterance context, prior utterance, and dialog context. This robust interpretation strategy is implemented by applying a set of mapping rules, which encode the contextual information stated above.

## 6 The Preliminary Evaluation

In the preliminary evaluation, we collected 122 recognized Chinese queries with microphone quality from the volunteer students. All of them are naïve users of spoken dialogue systems. Recognition performance is shown in Table 1. Currently, the queries are constrained to one frame type *routine*, i.e., the users are asked to communicate with the system and find the best traffic route between two locations in Shanghai city.

**Table 1.** Recognition error rates of the test set.

| | Num | % |
|---|---|---|
| Total queries | 122 | 100.0 |
| Queries with non-word errors | 68 | 55.7 |
| Queries with real-word errors | 7 | 5.7 |

**Table 2.** The comparative results of understanding experiments on the test set.

| | Understanding errors in simple framework (Num. | %) | Understanding errors in robust framework (Num. | %) |
|---|---|---|---|---|
| Clean queries | 5 | 4.1 | 4 | 3.3 |
| Corrupt queries | 62 | 50.8 | 33 | 27.0 |

We carried out a comparative understanding experiment on this test set. Our robust framework is compared with a simple semantic analysis solution, which is based on a standard chart parser and has no recognition repair module. All of the test sentences were labeled by hand with the semantic frames. A severe definition of understanding error is given as follows: an error occurs when there is at least one difference between the hypothesized frame and the reference frame. Table 2 reports the comparative understanding error rate of the two systems. From the evaluation, we can conclude that the proposed robust mechanisms improve effectively the understanding of the query, especially for the corrupt queries.

## 7 Conclusion

We have presented a framework for robust analysis and interpretation of spoken language, which integrates recognition error repair, robust semantic parsing, and robust pragmatic interpretation. The robust strategies can be outlined as follows:

- Combining the similarity-based and rule-based approaches for error correction;
- Allowing partial parsing and word skipping in the parser;
- Extensive use of high level knowledge for robust interpretation.

A preliminary evaluation showed that these robust processing strategies improve the understanding of the spoken Chinese queries effectively.

## References

1. Eckert, W., Niemann, H., Semantic Analysis in a Robust Spoken Dialog System. In Proc. Of ICSLP, Yokohama, 1994, pp.107-110.
2. W. Ward. Understanding Spontaneous Speech: The Phoenix System. In Proc. Of ICASSP 91, vol. 1, pp. 365-367.
3. C. J. Rupp and David Milward. A robust linguistic processing architecture. Siridus Report 4.1, Göteborg University, Department of Linguistics, Göteborg, Sep. 2000.
4. Satoshi Kaki, et al. A Method for Correcting Errors in Speech Recognition using the Statistical Features of Character Co-occurrence. COLING-ACL '98. Montreal, Canada, pp.653-657.

# Development and Control of a Hand Exoskeleton for Rehabilitation

Andreas Wege, Konstantin Kondak and Günter Hommel

Real-Time Systems and Robotics
Technische Universität Berlin, Germany
{awege, kondak, hommel}@cs.tu-berlin.de

**Abstract.** Hand injuries are a frequent problem. The great amount of hand injuries is not only a problem for the affected people but economic consequences follow because rehabilitation takes a long time. To improve therapy results and reduce cost of rehabilitation a hand exoskeleton was developed. For research on control algorithms and rehabilitation programs a prototype supporting all four degrees of freedom of one finger was build (s. Fig. 1). In view of the fact that a lot of hand injuries affect only one finger, this prototype could already be functional in physical therapy. A robust sliding mode controller was proposed for motion control of the hand exoskeleton. The performance of the proposed controller was verified in real experiments and compared to that of traditional PID controller.
**Keywords.** hand exoskeleton, rehabilitation, human-machine interaction, sliding mode control

## 1 Introduction

Hand injuries are a common result of accidents. Permanent impairments are regular consequences of these injuries. After hand operations it is essential to perform rehabilitation to regain previous dexterity. Social and economic consequences are severe if the result of rehabilitation is not optimal. As an example rehabilitation is necessary to treat flexor tendon injuries or to avoid scarring and adhesion after surgery. Another problem during the rehabilitation process is lack of reproducible measurements. These are needed to identify limitations in the dexterity of the hand and to evaluate the progress of rehabilitation.

Currently most rehabilitation is performed manually by physiotherapists. High personnel costs and lack of motivation to perform exercises at home present a problem. Some devices support physiotherapists by applying a continuous passive motion to the patient's hand, e.g. [1]. These devices are limited in the number of independently actuated degrees of freedom,



*G. Hommel and Sheng Huanye (eds.), Human Interaction with Machines,* 149-157.

and do not integrate sensors for diagnostics. The evaluation of progress is therefore done manually by the therapist. More flexible robotic support in rehabilitation of hand injuries is not common.

Research on hand exoskeletons is still going on but the majority of these devices were not developed with focus on rehabilitation. Other applications are haptic interaction with a virtual reality or remote manipulations with robot arms. Some hand exoskeletons restrict motions of the joints and do not exert them actively. An exoskeleton that uses ultrasonic clutches was presented in [2]. The commercially available *CyberGrasp from Immersion* restricts motion by pull cables with brakes on their distant end. Other devices can only exert forces in one direction. An example for this type of devices was presented in [3, 4]. For virtual reality these devices are suitable but for rehabilitation purposes a bidirectional active motion is desired. At the *Robotics Center-Ecole des Mines de Paris* a hand exoskeleton was developed that supports bidirectional movement for two fingers [5]. It supports up to four degrees of freedom per finger but controls only one of them at the same time through a pull cable. The *LRP Hand Master* is similar to the exoskeleton presented in this paper, but was not used for rehabilitation purposes yet [6]. It supports 14 bidirectional actuated degrees of freedom.

Recent research on the usage of exoskeletons in physical therapy was done by scientists of the University of Salford [7]. For their experiments they used a tendon driven exoskeleton which controls flexion of two degrees of freedom per finger. The *Rutgers Master II* actuates the fingers by four pneumatic pistons inside the palm. The device was employed in a study for the rehabilitation of stroke patients [8]. Another example for an exoskeleton used in rehabilitation showed the possibility to improve rehabilitation progress as well [9].

## 2 Construction of the Exoskeleton

Existing exoskeleton devices do not satisfy all needs of rehabilitation. The device should be modular, lightweight, and easy to attach even to deformed or scarred hands. Furthermore the palm should be free of any mechanical elements to allow interaction with the environment. Bidirectional movement in all degrees of freedom is desired. Each finger has four degrees of freedom: Flexion and extension in metacarpophalangeal (MCP) joint, proximal interphalangeal (PIP) joint, and distal interphalangeal (DIP) joint; and abduction/adduction in MCP joint.

The developed construction fulfils these requirements and supports bidirectional motion in all joints. It moves fingers by a construction of levers which are connected to the hand by orthopaedic attachments. Link lengths were designed to allow nearly full range of motion. Each lever ends in a pulley where two ends of a Bowden cable are fixed. The Bowden cables are connected to actuating motors with transmission gears. Flexible Bowden sheaths allow some movement of the hand in relation to the actuator unit. Movement of the cables within the sheaths leads to a rotation of the pulleys which results in a rotation of finger joints. Each pair of the Bowden sheaths are attached to a mount at the phalanxes. On the other end they are attached to a tension device to keep cables under tension reducing slackness.

On one joint of each lever a hall sensor is integrated. The corresponding finger joints angles can be calculated with help of trigonometric equations and known link lengths. The positions of motor axes are measured by optical encoders. These values are also used to calculate joint angles of the fingers. As a result of strain and slackness both values for joint angles deviate.

Force sensing resistors attached on top and on bottom of each phalanx measure the applied forces during flexion and extension. The resulting forces are inaccurate because not all forces are exchanged through the sensors. The contact area is maximized by applying distance pieces to assure that most of the force is passed through the force sensors. The values are accurate enough to measure dynamic changes in the applied forces. The actual currents of the motors can deviate from the values set by the control system. These currents and therefore the torques at the motor axes are measured as well. Together with transmission ratio of gears and leverage they can be used to estimate the force exerted to the phalanges. The friction of the Bowden cables has to be considered as a source of error.

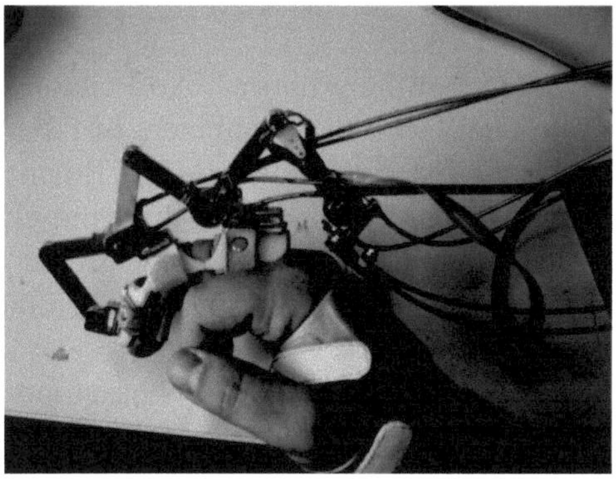

**Fig. 1.** Prototype of hand exoskeleton for one finger.

## 3 Control System

The control system consists of three parts: a real-time controller, the host computer running the interface for the instructor, and the client interface giving visual feedback to the user. The user and the instructor interface can run on separate computer if desired. The different parts of the system are connected through network interfaces.

The real-time controller from *National Instruments* is running the real-time operation system *Pharlap* and samples all sensor data (hall sensors, quadrature encoders, motor currents, and force sensors) through data acquisition cards. All control loops are executed on this controller as well. The motors are driven through analog output channels which are connected to PWM-controllers using a torque control mode. The system is designed for the control of the complete hand exoskeleton with 20 axes.

The host computer running the interface for the instructor allows the setting of desired motion and displays all sensor data. The system allows recording the motion of all joints. The motion can be filtered and replayed later at a customised speed. Remote assistance is possible as the application can run on any computer connected to the internet.

The control loops run independently for each controlled joint. The controlled variable for each control loop is the measured angle at the optical encoders of the motors. The measured angles from the hall sensors are not yet used inside the control loop. The redundant angle sensor data are currently only used to detect mechanical failures. A PID control and a sliding mode control were implemented. Experiments were performed to compare the performance of both controllers.

## 4 Control Algorithm

The precise motion control of the proposed exoskeleton is complicated because of the two following facts: the parameters of the mechanical system are changing by each trail due to the deflection of Bowden cables; the load of the actuators caused by the resistance of the hand changes. Therefore, the standard controllers like PID which can be well tuned only for some small operation area provide a bad performance when the load or parameters of the system are changed.

The proposed solution for motion control is based on application of the sliding mode control (SMC). One advantage of the SMC is the insensitivity to parameter variations and rejection of disturbances. As it was shown, see e.g. [10], a system, linear in the input

$$\dot{\mathbf{x}} = \mathbf{f}(\mathbf{x},t) + \mathbf{B}(\mathbf{x},t)u \tag{0.1}$$

can be forced to move in the state subspace, called sliding manifold, given by s(**x**)=0, where s(**x**) is a scalar function called sliding surface. To achieve the motion on the sliding manifold, discontinuous control is used:

$$u = \begin{cases} u_+(\mathbf{x}), & \text{for } s(\mathbf{x}) > 0 \\ u_-(\mathbf{x}), & \text{for } s(\mathbf{x}) < 0 \end{cases} \tag{0.2}$$

The control (0.2) works as follows: if the system state does not lie on the sliding manifold, the control $u$ takes one of its values $u_+$ or $u_-$ and steers the system state toward the sliding manifold. After reaching the sliding manifold the control performs infinite fast switching between $u_+$ and $u_-$ and the system state remains in the sliding manifold forever. If $u_+$ and $u_-$ can be made independent of system parameters, only the definition of the sliding surface determines motion of the system, and control (0.2) makes the system totally insensitive to parameter variations and disturbances.

In spite of its advantages, SMC is seldom used for control of robot motion. The main difficulty in connection with application of the SMC is the practical realization of infinite fast switching in the control (0.2) which leads to chattering (high frequency oscillation).

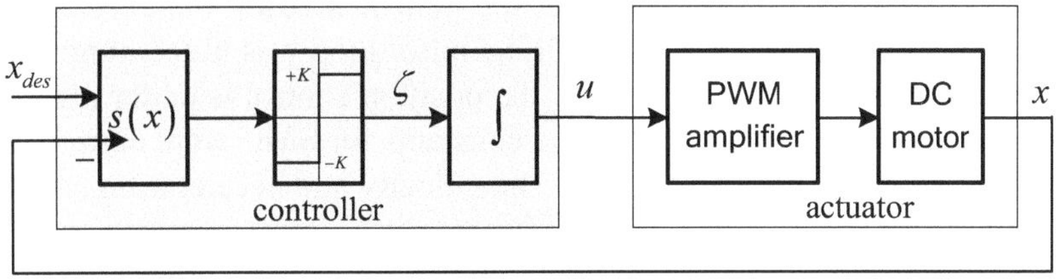


**Fig. 2.** The idea of the control schema.

To avoid the chattering problem the control scheme shown in Fig. 2 was proposed. An imaginary input $\zeta$ of the system is introduced and the real input $u$ will be considered as a part of the system state. The imaginary input $\zeta$ is integrated to form the real input $u$, which is then fed to the actuator. The integrator belongs to the controller and thus ideal switching for input $\zeta$ becomes possible. The resulting control rule for each actuator of the exoskeleton is therefore:

$$u = \int -K \operatorname{sign}\left(s\left(\mathbf{x}\right)\right) dt \tag{0.3}$$

The sliding surface was defined as follows:

$$s\left(\mathbf{x}\right) = C_q e + C_{\dot{q}} \dot{e} + C_{\ddot{q}} \ddot{e} = 0 \tag{0.4}$$

Here $e$ is the error in the coordinate and $C_i$ are constants. It can be easily shown, see e.g. [11], that if the system moves on sliding surface (0.4), the error progression can be forced to move along function exp(-$\lambda t$). For this case, the constants $C_i$ should be chosen as follows: $C_q$ is free, $C_{\dot{q}} = 2C_q / \lambda$ ,. $C_{\ddot{q}} = C_q / \lambda^2$ In [11] the stability of the proposed controller is proven for a large class of mechanical systems.

For time discrete implementation of the control law (0.3) the switching element should be approximated by means of linear function with a large slope. For a given controller period the slope of this linear approximation and the constant $K$ can be easily determined in an experiment.

## 5 Experimental Results

To evaluate the system, first experiments where performed with position control. The maximum torque of each motor gear combination is about four Nm. But for safety reasons the torques of the motors were limited to 20% of their maximum torque so that the human wearing the exoskeleton is stronger. The performance of the PID control loop was tuned at no load. for a stable motion. Less precision in the position control is better than undesired overshoots or oscillations. The constants for SMC were determined in experiments with no load as well. The velocity and acceleration are calculated by numerical differentiation. Both control loops currently runs at 2.5 KHz.

In the first experiment the performance was evaluated on a step response from 0 to 90 degrees in the MCP joint for both controller types. In that experiment the user was passive and not working with or against the motion. Fig. 3 shows the recorded data for the PID and sliding mode controller. The PID controller shows a satisfying performance. The deviation lies within one degree after reaching the desired value. But it could not be tuned for faster response maintaining the stability. The discontinuities while approaching the desired value are results from varying load and friction within the Bowden cables. In comparison SMC could be tuned for faster response maintaining the stability. The greatest deviation after

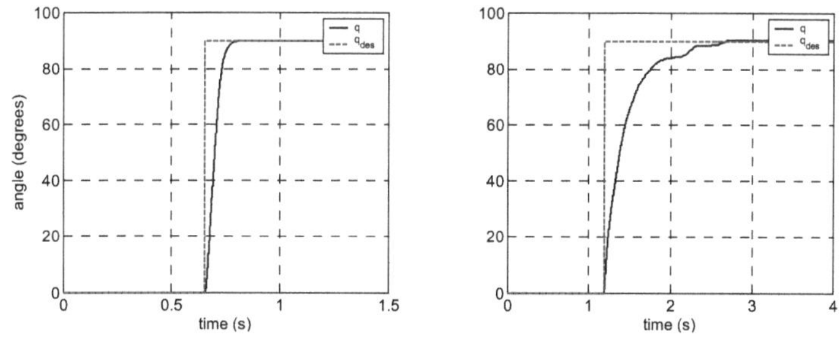


**Fig. 3.** Step responses of systems using SMC and PID control.

reaching the 90 degrees was within one step of the optical encoders. The trajectory is also smoother compared to the PID controller.

In a second experiment the controllers were set to hold a constant position. The human wearing the hand exoskeleton varied the applied force. The PID controller could stabilize the position within one degree for a slowly varying force (Fig. 4 left). The sliding mode controller performed better even with greater forces applied to the exoskeleton (Fig. 4 right). The greatest deviation in the position was within three encoder steps. The motor torques changing with high frequency occurred due to deflections in the Bowden cable. However these rapid changes did not result in undesired vibrations at the finger joints.

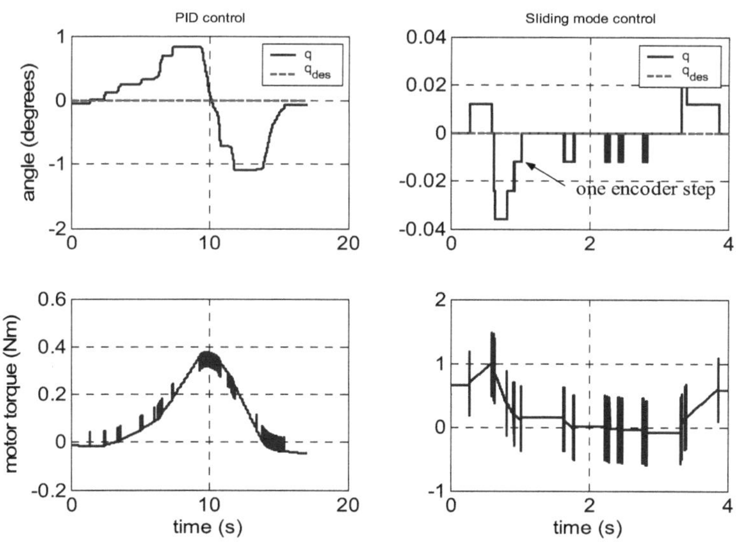


**Fig. 4.** Holding constant position under varying force with PID control and SMC.

## 6 Conclusions and Future Work

The presented work is a basis for future research and clinical studies. The hand exoskeleton was developed under consideration of the special needs of the rehabilitation. Nearly every possible trajectory of motion can be applied. The sliding mode control allows a fast and stable position control. The integrated sensors allow measurements for new methods in rehabilitation and diagnostic of hand injuries. Automatic adaptation to the progress of the rehabilitation for individual patients becomes possible (e.g. by measuring their resistance to the applied motion).

Next steps will include the assembly of the hand exoskeleton to support all four fingers and the thumb. Calibrated force sensors will allow a better force measurement. The integration of an automatic calibration of the finger joint angles could simplify the usage of the exoskeleton as well [12].

The position accuracy of the control loop could be further improved by integrating the measured angles from the hall sensors. By that measure the tolerance introduced by the flexibility of the Bowden cables could be reduced. Further control modes incorporating the measured force will allow more flexible training programs. As an example the speed of the trajectory could be varied according to the resistance of the patient against the movement which is measured by the force sensors. This would allow the system to synchronize with the patient and avoid forcing the timing of the exercises onto the patient. The comfort of the rehabilitation program and possibly the success could be improved by this way.

Another desired control mode is a force control mode where the user is solely defining the motion by his or her movements; friction introduced by the exoskeleton will be suppressed. In this mode measurements are possible without interference of the hand exoskeleton with the patient's movements.

## References

1. JACE H440 continuous passive motion device for fingers: http://www.jacesystems.com/products/hand_h440.htm.
2. Koyama, T.; Yamano, I.; Takermua, K.; Maeno, T; Development of an Ultrasonic Clutch for Multi-Fingered Exoskeleton using Passive Force Feedback for Dexterous Teleoperation, Proceedings of the 2003 IEEE/RSJ International Conference on Intelligent Robots and Systems, 2003, Las Vegas Nevada.
3. Avizanno, C. A.; Barbagli, F.; Frisoli, A.; Bergamasco, M.: The Hand Force Feedback: Analysis and Control of a Haptic Device for the Human Hand, Proc. of IEEE, Int.. Conf. on Systems, Man and Cybernetics, 2000, Nashville.

4. Avizzano, C. A.; Marcheschi, S.; Angerilli, M.; Fotana, M.; Bergamasco, M.: A Multi-Finger Haptic Interface for Visually Impeared People, Proceedings of the IEEE International Workshop on Robot and Human Interactive Communication, 2003, Millbrae, California, USA .
5. Stergiopoulos, P.; Fuchs, P. and, Laurgeau, C.: Design of a 2-Finger Hand Exoskeleton for VR Grasping Simulation, EuroHaptics 2003.
6. Tzafestas, C.S.: Whole-Hand Kinestethic Feedback and Haptic Perception in Dextrous Virtual Manipulation, IEEE Transactions on Systems, Man, and Cybernetics - Part A: Systems and Humans, Vol. 33, No. 1, 2003.
7. Sarakoglou, I.; Tsagarakis, N.G.; Caldwell, D.G.: Occupational and physical therapy Using a hand exoskeleton based exerciser, Proc. of 2004 IEEE/RSJ International Conf. on Intelligent Robots and Systems, 2004, Sendai, Japan.
8. Jack D.; Boian R.; Merians A.; Tremaine M.; Burdea G.; Adamovich S.; Recce M.; Poizner, M.: Virtual Reality-Enhanced Stroke Rehabilitation In: IEEE Transactions on Neural Systems and Rehabilitation Engineering, Vol. 9, No. 3, September 2001.
9. Brown, P.; Jones D.; Singh S.; "The Exoskeleton Glove for Control of Paralyzed Hands"; in IEEE 1993.
10. V. I. Utkin. Sliding Modes in Control and Optimization. Springer 1992. ISBN 3-540-53516-0.
11. K. Kondak, G. Hommel, B. Stanczyk, M. Buss. "Robust Motion Control for Robotic Systems Using Sliding Mode". Proposed for IEEE/RSJ Int. Conf. on Intelligent Robots and Systems 2005.
12. A. Wege, K. Kondak and G. Hommel: "Self-Calibrating Joint Angle Measurements For Human Fingers Using Accelerometer And Gyroscope Sensors", International IEEE Conference on Mechatronics & Robotics, Aachen 2004.

# Automatically Constructing Finite State Cascades for Chinese Named Entity Identification

Tianfang Yao

Department of Computer Science and Engineering
Shanghai Jiao Tong University
Shanghai 200030, China
yao-tf@cs.sjtu.edu.cn

**Abstract.** In this paper, we propose a new approach that can automatically construct Finite State Cascades (FSC) from given sets of rules applied to Chinese named entity (NE) identification. In this approach, the regular expressions employed in FSC are permitted to represent complex constraints. Thus, it extends the original definition of Finite State Automaton (FSA) and makes FSC more suitable to real-world applications. Additionally, the construction procedure of FSC is transparent for NE recognition rule developers. Therefore, FSC is more flexible and maintainable. Moreover, it can easily be adapted to other domains because of the independence from domains. The experimental result has shown that total average recall, precision, and F-measure for the identification of three Chinese named entities have achieved 87.20%, 84.20%, and 85.65% respectively. It shows that the extended definition of FSC is sound and the algorithm used for the automatic construction of FSC is appropriate and effective.

## 1 Introduction

The task of information extraction is to find and extract relevant information from large volumes of free texts with the aim of filling database records in an efficient and robust way. In the investigation of Chinese named entity (NE) identification, we adopt *football competition news* as our corpus, because a variety of NEs exist in the news. Among the NEs, we select six of them as the recognized objects, i.e., *personal name* (*PN*), *date or time* (*DT*), *location name* (*LN*), *team name* (*TN*), *competition title* (*CT*) and *personal identity* (*PI*) (Yao et al. 2003).

Although there are different approaches to identifying Chinese NEs (Chen et al. 1998; Chen and Chen 2000; Zhang and Zhou 2000; Sun et al. 2002), considering Chinese NE's construction as well as the comprehensive superiority in *accuracy*, *efficiency*, and *robustness* of identification,



*G. Hommel and Sheng Huanye (eds.), Human Interaction with Machines,* 159-165.

we utilize *Finite-State Cascades* (*FSC*) (Abney 1996) as a shallow parser to identify them.

Generally speaking, FSC translated by regular expressions are constructed by standard compiler techniques (Abney 1997; Zhang 1998) or other techniques (Piskorski et al. 2002). These expressions, however, merely include a single constraint symbol[1] such as Part-of-Speech (POS) constraint. If regular expressions embody complex constraint symbols, e.g., both POS and semantic constraint symbols, it is difficult to adopt the above techniques to construct FSC. In our recognition rules, sometimes a POS constraint symbol may correspond to multiple semantic constraint symbols in a rule. Therefore, we extend the original definition of Finite-State Automata (FSA). With this extension, we can improve the *practicability* of FSC mechanism. Based on the extension, we propose an approach to *automatically* construct FSC using NE recognition rule sets for conveniently constructing and maintaining FSC.

## 2 Automatically Constructed Finite-State Cascades

### 2.1 Definition of Recognition Rules

The representation of NE recognition rules are defined as follows:

Recognition Category → POS Constraint | Semantic Constraint$_1$ | Semantic Constraint$_2$ | … | Semantic Constraint$_n$

The left-hand side (LHS) of the rule is a recognition category that indicates a recognized NE; while the right-hand side (RHS) of the rule lists POS constraint and its corresponding one or more semantic constraints. Meanwhile, the symbol "|" denotes a separation between constraints. Additionally, the symbol "→" between LHS and RHS represents a productive (or conventional) relationship of these two sides. The rule tag set contains 19 POS and 29 semantic tags. A POS constraint describes the part-of-speech and sequence of named entity's constituents. In addition, a semantic constraint gives the meaning of corresponding part-of-speech or constituents, such as country name, province (state) name, city name, company name, club name, and product name, etc. The following example specifies a rule for TN identification.

---

[1] The single constraint symbol means that the symbol in a regular expression is an atomic symbol.

***Example:***

TN → N + KEY_WORD | AbbreviationName + TeamNameKeyword | CityName + TeamNameKeyword | CompanyName + TeamNameKeyword | ClubName + TeamNameKeyword | CountryName + TeamNameKeyword | ProductName + TeamNameKeyword | TNOtherName + TeamNameKeyword

This rule means that N (noun) must be an abbreviation name, a city name, a company name, a club name, country name, a product name, or another team name. Following N, a KEY_WORD must occur. Notice that KEY_WORD represents the trigger word within a team name, such as the words 队 (Team), 联队 (League), etc.

## 2.2 Construction Algorithm

For clearly expressing some fundamental concepts used in our construction algorithm, we give the following definitions. Notice that they extend the definitions given by Piskorski (2002).

**Definition 1.** A finite-state automaton (FSA) is a 6-tuple $M = (Q, S_1, S_2, \delta, i, F)$, where Q is a finite set of states; $i \in Q$ is the initial state; $F \subseteq Q$ is a set of final states; $S_1$ and $S_2$ are two finite character string sets; and $\delta: Q \times S_1 \times S_2 \rightarrow 2^Q$ is the transition function. Moreover, δ is extended to $\delta': Q \times S_1^* \times S_2^* \rightarrow 2^Q$ for accepting strings over $S_1 \times S_2$. The language accepted by M is defined as L(M):

$$L(M) = \{(u, v) \in S_1^* \times S_2^* \mid \delta'(i, (u, v)) \cap F \neq \emptyset\} \quad (2.1)$$

**Definition 2.** Finite-state cascades (FSC) is a $n$-tuple $NM = (M_1, M_2, \ldots, M_n)$, where $n$ is a level number of FSC. $M_1, M_2, \ldots, M_n$ are $n$ FSAs, in which $M_1$ and $M_n$ correspond to the lowest-level and highest-level FSAs of FSC respectively.

**Definition 3.** The automatic construction of FSC is to automatically transform ordinal recognition rule sets into NM, that is, each rule set is transformed into a $M_i$ from low to high level.

According to the *graph theory* (Aho et al. 1983), we proposed an algorithm for automatically constructing FSC. The main ideas concerning optimized construction are: (i) under the condition of correct construction, the redundancy of edges and states should be *minimized*; (ii) for the sake of reducing the complexity of FSC, the self-edges that begin and end at the same state and the cycles between two states are *not* considered. Based on such a construction strategy, we can ensure the *correctness* of FSC and also enhance the *efficiency* of NE identification.

**Definition 4.** In the construction algorithm, in order to effectively store nodes and edges of FSC, we use four adjacent matrices, *POS matrix*

(*POSM*), *POS index matrix* (*POSIM*), *semantic index matrix* (*SIM*) and *semantic constraint matrix* (*SCM*), which are used as data structure, to replace the transition function δ:

- POSM is used to store POS tags from recognition rules between two states: POSM = {($q_1$, $q_2$, $S_1$) | $q_1$, $q_2 \in Q$; $S_1 \subseteq S$; $q_1$ is a starting state, $q_2$ is an arriving state, $S_1$ is a set of all accepted POS tags between $q_1$ and $q_2$, and S is the set of all POS tags.}
- POSIM provides the line address pointer to SIM, which is related to POS tags between two states: POSIM = {($q_1$, $q_2$, $la_{SIM}$) | $q_1$, $q_2 \in Q$; $la_{SIM} \in LA_{SIM}$; $q_1$ is a starting state, $q_2$ is an arriving state, $la_{SIM}$ is a line address of SIM; $LA_{SIM}$ is the set of all line addresses of SIM.}
- SIM indicates the position of semantic tags associated with each POS tag in SCM: SIM = {($la_{SIM}$, $s_1$, $la_{SCM}$) | $la_{SIM} \in LA_{SIM}$; $la_{SCM} \in LA_{SCM}$; $s_1 \in S_1$; $la_{SCM}$ is a line address of SCM; $LA_{SCM}$ is the set of all line addresses of SCM; $s_1$ is a POS tag of S.}
- SCM saves the semantic tags for each POS tag in POSM: SCM = {($la_{SCM}$, $s_2$, bool) | $la_{SCM} \in LA_{SCM}$; $s_2 \in S_2$; bool = 1; $s_2$ is one of the accepted semantic tags corresponding to $s_1$ in SIM.}

To distinctly describe FSC construction procedure, we give the following primary construction algorithm by pseudo codes:

### *Construction Algorithm:*

```
main()

input level_size;
for (level_index = 1 to level_size)
        initialize recognition_rule();
        input a NE recognition rule set into recognition_rule();
        initialize posm(), posim(), sim(), and scm();
        state_index ← 0;
        initial_state ← state_index;
        rule_set_size ← get_rule_size(recognition_rule());
        for (rule_index = 1 to rule_set_size)
                {get a recognition rule containing POS and SEM tags}
                pos_rule() ← recognition_rule(rule_index, POS);
                sem_rule() ← recognition_rule(rule_index, SEM);
                if (rule_index == 1)
                        {add the first recognition rule in the recognizer}
                        pos_tag_size ← get_tag_size(pos_rule());
                        for (pos_tag_index = 1 to pos_tag_size)
                                pos_tag ← pos_rule(rule_index, pos_tag_index);
                                add pos_tag into posm(state_index, state_index+1);
```

```
                add sem_rule(rule_index, pos_tag) into scm() indexed
                by posim() and sim();
                state_index ← state_index + 1;
            final_state ← state_index;
    else
            {add other recognition rules in the recognizer }
            add_rule(pos_rule(), sem_rule(), rule_index , initial_state,
            final_state)²;
{put a constructed recognizer into a level of the FSC}
fsc_posm(level_index, , ) ← posm();
fsc_posim(level_index, , ) ← posim();
fsc_sim(level_index, , ) ← sim();
fsc_scm(level_index, , ) ← scm();
```

Notice that it is important that the *correct construction condition* in the procedure of adding a new POS tag's edge must be met. For example, whether its corresponding semantic tags *conflict* with the semantic tags of an existing POS tag edge in a NE recognizer. Otherwise, adding this edge is *given up*.

Obviously, the construction algorithm is a *rule-driven* algorithm and only relies upon the representation of rules. Therefore, the constructed FSC are flexible and maintainable in that it is *easy* to change the size of POS and semantic tags, and *easy* to add, modify or delete recognition rules. Additionally, because this algorithm can be applied to establish different recognition levels, it is also *easy* to expand the NE recognizers in FSC for *new NEs*.

## 3 Experimental Results

The above algorithm has been implemented in Java in the prototype system *CHINERIS (Chinese Named Entity and Relation Identification System)*. The system can automatically identify 6 types of NEs and instances of 14 NERs in the domain of football matches.

In the experiments, we *randomly* chose 20 texts including 658 sentences (roughly 8340 characters) from Jie Fang Daily in May 2002 to evaluate the named entity identification. Notice that a gazetteer is used for identifying static named entities such as continent, country, province (state), city, club, company, product name, etc., before TN, CT and PI are recognized, because these static named entities are the constituents of TN and CT. Additionally, because FSC provide a multi-level recognition mechanism, we

[2] Because of space limitation, we omit the pseudo codes for add_rule().

arrange a recognition rule set in the corresponding NE recognizer. Furthermore, the rule sets provided for TN, CT, and PI recognition have 35, 50, and 20 rules respectively. Considering the sequence of NE identification (e.g., in PI recognition rules TN and CT categories are used as POS constraints), we put TN, CT and PI recognizers in low to high order (Another three NEs, namely, PN, DT and LN are immediately recognized after error repair for word segmentation and POS tagging).

Three measures, i.e., recall, precision, and F-measure, are applied to evaluate the performance of NE identification. The result is shown in the Table 1. Positive result proves that the extended definition of FSC is *sound* and the algorithm used for the automatic construction of FSC is *correct* and *effective*. Moreover, this framework can easily be adapted to other domains because of the *independence* from domains.

**Table 1**. Performance for TN, CT and PI Identified by FSC.

| | TN | CT | PI | Total Average |
|---|---|---|---|---|
| Recall | 81.95 | 86.65 | 93.00 | 87.20 |
| Precision | 76.35 | 87.65 | 88.60 | 84.20 |
| F-measure | 79.05 | 87.15 | 90.75 | 85.65 |

## 4 Conclusion

In this paper, an approach to automatically construct FSC for Chinese NE identification has been presented. Compared to the original definition of FSA, our regular expressions used to construct FSC are allowed to represent *complex* constraints. Thus, it makes FSC more suitable to *practical applications*. Moreover, because the parser is based on automatically constructed FSC, it is more *flexible* and *maintainable*. Finally, it is *easy* to use the FSC-based parser for rule developers. They don't need to concern FSC's construction modes, that is to say, it is *transparent* for them how to construct FSC. The positive experimental result proves that the extended definition of FSC is *sound* and the algorithm used for the automatic construction of FSC is *appropriate* and *effective*.

## Acknowledgement

This work is a part of the COLLATE project under contract no. 01INA01B, which is supported by the German Ministry for Education and Research.

## References

1. Abney S (1996) Partial parsing via Finite-State Cascades. Proc of the ESSLLI '96 Robust Parsing Workshop:8-15.
2. Abney S (1997) Part-of-Speech tagging and partial parsing. In: Young S and Bloothooft G (eds) Corpus-based methods in language and speech processing, Kluwer Academic Publishers, Dordrecht, pp. 118-136.
3. Aho A, Hopcroft J, Ullman J (1983) Data structures and algorithms. Addison-Wesley Publishing Company, Massachusetts California London Amsterdam Ontario Sydney.
4. Chen H, Ding Y, Tsai S, Bian G (1998) Description of the NTU system used for MET2. Proc of the seventh Message Understanding Conference (MUC-7).
5. Chen K, Chen C (2000) Knowledge extraction for identification of Chinese organization names. Proc of the second Chinese processing workshop:15-21.
6. Piskorski J (2002) The DFKI finite-state machine toolkit. Research Report RR-02-04, DFKI, Saarbrücken, Germany.
7. Piskorski J, Drożdżyński W, Xu F, Scherf O (2002) A flexible XML-based regular compiler for creation and conversion of linguistic resources. Proc of the third international conference on language resources and evaluation (LREC-2002).
8. Sun J, Gao J, Zhang L, Zhou M, Huang C (2002) Chinese named entity identification using class-based language model. Proc of the 19th international conference on computational linguistics (COLING 2002):967-973.
9. Yao T, Ding W, Erbach G (2003) CHINERS: A Chinese named entity recognition system for the sports domain. Proc of the second SIGHAN workshop on Chinese language processing (ACL 2003 Workshop):55-62.
10. Zhang Y (1998) Chinese text interpretation based on hybrid method. PhD thesis, Shanghai Jiao Tong University, Shanghai, China.
11. Zhang Y, Zhou J (2000) A trainable method for extracting Chinese entity names and their relations. Proc of the second Chinese language processing workshop (ACL 2000 Workshop):66-72.

# Improving Information Retrieval by Concept-Based Ranking

Martin Mehlitz, Fang Li

Dept. of Computer Science, Shanghai Jiao Tong University

**Abstract.** With the Internet getting available to more and more people in the last decade and with the rapidly growing number of webpages the internet is a vast resource of information. Millions of people are using the internet to search for information every day. However, the search result is not satisfying in the case of an ambiguous search query. In this paper an algorithm for re-ranking pages according to concepts - the content of webpages – is proposed. This algorithm uses Association rules, a Data Mining technique, to derive concepts from webpages. The preliminary experiment shows a promising result compared with hyperlink-based ranking algorithms.

## 1 Introduction

By the end of February 2003 more than 600 million searches on the web were performed as a daily average[1]. Many users use very simple search queries that often return a lot of irrelevant pages. Almost all these searches are performed by web services like google, which uses the Page score [2], which scores a page according to the following policy: A webpage is important, if many other important webpages have a references to this page. This is a ranking approach which disregards the content of the webpages. How to rank pages based on concepts is the main task of improving the precision of information retrieval. We propose a method to derive concepts from a set of webpages. Re-ranking pages according to a concept will greatly reduce the number of irrelevant pages that are viewed. The concepts are derived automatically from the webpages, they consists of words that appear together frequently. In the following, we first describe some related work, then the concept-based ranking algorithm and finally a preliminary result is reported.

[1] From www.searchengine.com



*G. Hommel and Sheng Huanye (eds.), Human Interaction with Machines,* 167–176

## 2 Related Work

People searching for information on the web are called informavores [5] and following this analogy their search for information is compared to the search for food performed by animals. Based on the former information consuming behaviour shown by the user, the spreading activations cost theory from biology is applied to predict whether a page is interesting to a user or not. An automatic resource compiler is introduced in [4] where - given a topic - a list of good hubs and good authorities will be compiled. The link structure of the webpages relevant to this topic is analysed and the anchor text on the pages is searched for words which occur in the description of the topic. [3] introduces a machine learning approach for re-ranking pages. The approach shows very good results but it needs the user to provide a sample page which's features are a positive sample for the machine learning part of their program. It can be very cumbersome if one has no such page and would first have to search the web for one. Our method also involves a user in the re-ranking process but the user will not be required to provide a page.

## 3 Concept-based ranking

### 3.1 Architecture

Figure 1 shows the architecture of our system. A standard scenario looks like this: A user enters a query, and then the pages returned by the google search engine are retrieved. These pages will be pre-processed, then the keywords will be extracted and the frequent item sets will be derived. In the next step, our algorithm for building concepts will create the concept item sets from the frequent item sets. These will be shown to the user who chooses a concept which will be used to re-rank the pages.

#### *The page retrieval*

The google search engine will be called to get the addresses of the first 100 pages for the given search string. For the search options the language restriction is set to English and google filtering is enabled, thus duplicate web pages won't be considered. Both .pdf and .doc files will be filtered out. Once we got the addresses we download all the pages and give them to the page pre-processing.

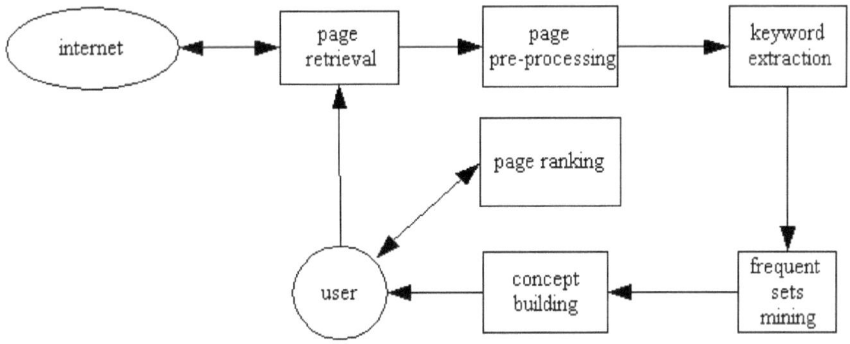


**Fig. 1.** Architecture.

### *Page pre-processing*

In pre-processing first the entire html-tags and hyper text relevant information are stripped from the pages, so that only the plain text which will be visible in a web browser remains. Then common stop words will be removed from a page as well as some words that frequently appear in web pages and are not related to the displayed information (like: www, webpage, forum, copyright, tel, fax ...). Occurrences of the words from the search string will also be removed.

### *Keyword extraction*

The very simple term frequency measurement is used for the keyword extraction. That is, a word that appears more frequent on a webpage than other words is considered more important than other words. Every word ***w*** on a page ***p*** will get a score according to the following equation:

$$\text{Score}(\boldsymbol{w})=\log(1+(\boldsymbol{n}^*\boldsymbol{length})/\boldsymbol{maxLength}) \quad (1)$$

Where ***length*** is the length of ***p*** and ***maxLenght*** is the length of the longest document. In our preliminary experiments the best results were achieved with taking the fifteen highest ranked words for each page as a keyword item set.

## 3.2 Mining frequent item sets

The Apriori algorithm [1] is used to mine frequent item sets from the keyword item sets. We set a fixed value of 0.1 (or 10%) as the minimum support for frequent item set. This means only words that occur together in at least ten pages will be considered frequent. A maximum support value is set to 0.5 (or 50%) in order to filter out co-occurrences which don't belong to a concept but to the structure of the web pages. The preliminary experiments showed that such a threshold is necessary because the web pages run only through very simple pre-processing. The output of this step is three sets of item sets (the 2-itemsets, the 3-itemsets and the 4-itemsets). Item sets of higher cardinality are not needed for the algorithm to work.

For each item set the confidence value is computed. Usually, confidence is a measure for rules of the form A=>B, defined as

$$\text{Confidence (A=>B)} = P(B|A) \quad (2)$$

This is, the percentage of transaction which include B given that they include A. We define confidence for our item set as the percentage of transaction that include all items of an item set given that the transaction includes any subset of these items.

## 3.3 Forming Concepts

We assume that pages about a certain concept have a more or less unique set of words that occur together more frequently than in pages that deal with other concepts. For example: pages about traveling to Shanghai will have the names of site-seeing spots like *Jin Mao Tower* or *Nan Jing Lu* appear frequently as well as words that have something to do with traveling like *hotel*, *flight* or *tickets*. But usually, none of the pages will have all of these words appear in them, nor will always the same words appear together. So we define those words that appear together with a very high confidence to be the *seeds* of the concepts. These *seeds* will be expanded with those item sets which have a slightly lower confidence and which have overlap with a *seed*. The latter item sets will be named *sprouts* and the confidence cut-off values which define *seeds* and *sprouts* are the minimum seed confidence (***msec***) and the minimum sprout confidence (***msprc***).

The algorithm for building the concepts from the frequent item sets works as follows:

Let $\boldsymbol{I}$ be the output of the above step. Then the output $\boldsymbol{\Omega}$ of the algorithm is a set of item sets which are the concepts occurring in the pages.

1. Initialize ***Ω*** to the empty set.
2. For each ***i*** in ***I***
    - if confidence of ***i*** > ***msec***
        - remove ***i*** from ***I***
        - put it into ***Ω***
3. if ***Ω*** is still empty
    - take the element with the highest confidence value out of ***I*** and put it into ***Ω***
4. for each ***k*** in {2,3, 4}
    - for each ***o*** in ***Ω***
        - for each ***i*** in ***I*** with confidence greater than ***msprc***
            - if (***k***-1) items in ***i*** occur also in ***o***
                - put the item that ***i*** and ***o*** do not have in common in a set ***t***
        - add ***t*** to ***o***
5. merge similar concepts
6. return ***Ω***

In step 1, the set of concepts will be initialized to the empty set. Then in step 2, the *seeds* are generated, these are those frequent item sets which have a very high confidence, representing those keywords, which almost always occur together. A fixed value (for example, 75%) will be applied. Step 3 is just for making sure that there will be at least one concept, by taking the item set with the highest confidence value as a *seed*. Then in step 4 the *seeds* will be expanded with there *sprouts*. A *sprout* to a *seed* is defined as follows:

Be $\boldsymbol{i}_{seed}$ a *seed* and $\boldsymbol{i}_{spr}$ a k-item set. If exactly k-1 items in $\boldsymbol{i}_{spr}$ are also elements of $\boldsymbol{i}_{seed}$ and if confidence ($\boldsymbol{i}_{spr}$)>***msprc***, then $\boldsymbol{i}_{spr}$ is a *sprout* of $\boldsymbol{i}_{seed}$.

Finally, before returning the concept item sets, all those sets with a very high overlap will be merged together. Overlapping sets are created because for one concept there might be more than one *seed*.

After the user chose a concept, the pages will be re-ranked according to how many words of the concept item set are present and how important these words are for a page using term frequency.

# 4 Experiments

## 4.1 Training parameters

In two experiments the optimal values for *msec* and *msprc* were determined. The sets which are produced differ greatly with these values. If the confidence values are too low, different concepts might be merged together. If the values are too high, the concept item sets will lack words that are relevant for a concept. We defined two different scenarios where a user is searching for information on the web. The first 100 pages from the google search results were downloaded and categorized according to the concept which they belong to. We need the labelled webpages only for the training phase to determine the optimal values for *msec* and *msprc*; they are not necessary for the algorithm to work. The integral of the graph that shows the relevant pages over the over-all viewed pages is our measure for how good a ranking of pages is for a certain concept (see Fig. 3-5).

**Table 1.** Results of confidence value tuning.

| | 25% | 30% | 35% | 40% | 45% | 50% | 55% | 60% | 65% | 70% | 75% |
|---|---|---|---|---|---|---|---|---|---|---|---|
| 25% | 0,56 | 0,56 | 0,56 | 0,56 | 0,56 | 0,56 | 0,56 | 0,85 | 0,85 | 0,53 | 0,53 |
| 30% | 0,82 | 0,82 | 0,56 | 0,56 | 0,56 | 0,82 | 0,85 | 0,85 | **0,85** | 0,52 | 0,52 |
| 35% | 0,83 | 0,83 | 0,83 | 0,76 | 0,81 | 0,84 | 0,85 | 0,84 | 0,84 | 0,52 | 0,52 |
| 40% | 0,84 | 0,84 | 0,84 | 0,84 | 0,84 | 0,85 | 0,85 | 0,84 | 0,84 | 0,52 | 0,52 |
| 45% | 0,83 | 0,83 | 0,83 | 0,83 | 0,83 | 0,85 | 0,85 | 0,84 | 0,84 | 0,52 | 0,52 |
| 50% | 0,84 | 0,84 | 0,84 | 0,84 | 0,84 | 0,84 | 0,84 | 0,84 | 0,84 | 0,52 | 0,52 |
| 55% | 0,84 | 0,84 | 0,84 | 0,84 | 0,84 | 0,84 | 0,84 | 0,84 | 0,84 | 0,51 | 0,51 |
| 60% | 0,78 | 0,78 | 0,78 | 0,78 | 0,78 | 0,78 | 0,78 | 0,78 | 0,78 | 0,51 | 0,51 |
| 65% | 0,78 | 0,78 | 0,78 | 0,78 | 0,78 | 0,78 | 0,78 | 0,78 | 0,78 | 0,52 | 0,52 |
| 70% | 0,52 | 0,52 | 0,52 | 0,52 | 0,52 | 0,52 | 0,52 | 0,52 | 0,52 | 0,52 | 0,52 |
| 75% | 0,52 | 0,52 | 0,52 | 0,52 | 0,52 | 0,52 | 0,52 | 0,52 | 0,52 | 0,52 | 0,52 |

Rows are the seed confidence – columns are the sprout confidence

In steps of 5 % all combinations of values for *msec* and *msprc* are tested. For each combination of values the concept item sets are built. The pages are then ranked with each item set and for each concept (the concepts found during the hand-labelling process) the best item set is selected. The sum of these values is the score for the *msec*/*msprc* combination. Table 1 is an excerpt from the normed (and rounded) values that were obtained from one of the scenarios. They are almost identical to the results in the other scenario and we found an msec value of 65% and an msprc value of 30 % are the best values for the first scenario and a combination of 50% and 40% for the second scenario.

## 4.2 Scenario 1

In this scenario a user searches for information about the song with the title *American Pie*, maybe searching for the lyrics or some general information like who is the interpreter, or when the song was released. Since recently three Hollywood movies with similar names have been released there will be a lot of pages that are reviews, promotion and so on for this movie. So there are obviously at least the concept of *movie* and the concept of *song* among the pages returned to a search string on a search engine.

Among the first 100 results returned for the search string "American Pie" 57 webpages are related to the movies with this name, 30 webpages are related to the song with this name, 7 belong to diverse other topics and 6 pages were not reachable. Figure 2 (left side) shows the best values for the minimum seed and sprout confidence we found, as well as all the concept item sets which were the output of the algorithm.

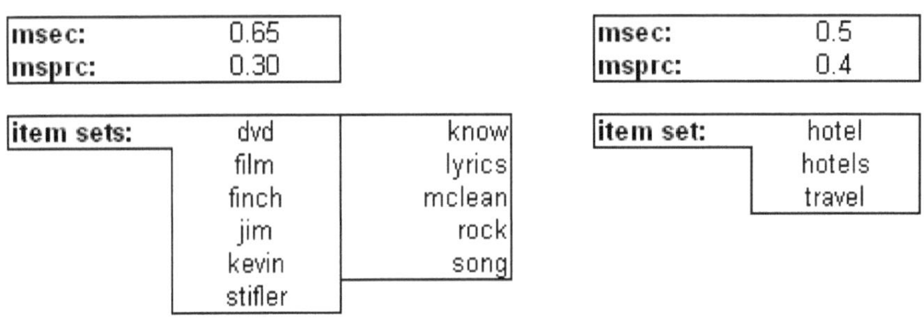


**Fig. 2.** Concept item set for scenario 1 (left) and scenario 2 (right).

### 4.3 Scenario 2

In this scenario a user is searching for some information about travelling to Shanghai. He plans a trip to Shanghai and want to know the cost of flights, hotel rooms, site seeing tickets and so forth. He chooses "china shanghai" as the search string. The first 100 results returned for the search string contain 22 webpages which are related to the concept of *travelling* (pages about booking hotels, flights or complete trips, or pages where china travellers report about there experiences) and 62 pages that relate to diverse other topics, the Formula 1, companies in Shanghai, consulates, studying and so on, but none of these concepts has more than three to five pages. Five webpages of the first 100 results could not be obtained.

This example is very "noisy", only one concept is present with more than the required number of pages. This makes it quite easy to filter out the pages which belong to this one concept, but if a user is interested in finding pages to the concepts that our algorithm ignores because they are not present in enough pages, the user will not find the concept at all. Figure 2 (right side) shows the best values for the minimum seed and sprout confidence we found, along with the concept item set which was the output of the algorithm.

## 5 Conclusion

In the conducted experiments concept item sets were derived from webpages. In two scenarios the item sets contain only concepts relevant words. Words like for example *song*, *lyrics* and *rock* appear together frequently and form a concept of *song*. We do not try to label the concepts, since it is a very difficult task of natural language understanding - someone who is searching for information about a topic will have some kind of comprehension about that topic.

Figures 3-5 show the ranking graphs for the three concepts we found in the two scenarios. Each figure shows the graphs for the best/worst possible ranking, for the google ranking and for the ranking according to the item sets we found. The *optimum* graph is the one that represent an ordering of the webpages, where all the relevant pages are viewed first, while the *worst* graph represents an ordering of the webpages, where all the irrelevant pages are viewed first. For each concept after re-ranking more concept-relevant are viewed earlier than for the google ranking.

In the Scenario *American Pie* ten pages among the first ten and still nineteen pages among the first twenty pages are all related pages. Compared to the google ranking with two concept-related pages among the top ten and nine pages among the first twenty, this is a significant improvement. Both of the other concepts (*movie* and *travel*) show the same tendencies.

We demonstrated the usefulness of the proposed algorithm in two experiments, and showed that association rules a - technique from Data Mining - can be used to improve Information Retrieval. The work of this paper will be the baseline for our future work.

The experiments showed that the crucial factor for getting useful concept item sets is the configuration of the values for the minimum seed and sprout confidence. Even though all the experiments show similar tendencies for the optimal values, they do differ, so simply choosing a fixed value for them might reduce the improvement that is achieved for some cases, so we plan to do further research on finding a generic method for determining the optimal values for *msec* and *msprc*, from the unlabeled keyword sets.

In this paper the computational cost was not addressed. The future goal of this research is a web application that computes in real-time with limited resources, where the computational cost is an important issue. Presently, the bottleneck for the program lies in obtaining and pre-processing webpages. A possible solution is to use the snipplets returned by the google search engine instead of the whole webpage and we will also do some research if this proves to be a good solution.

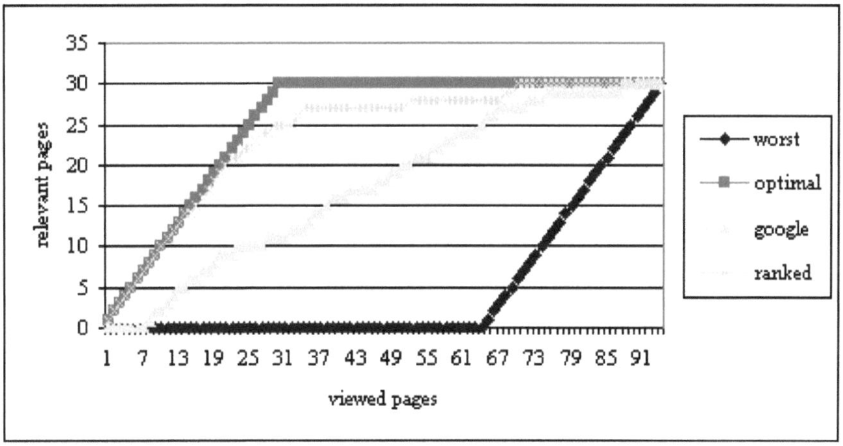


**Fig. 3.** Ranking for the concept *song.*

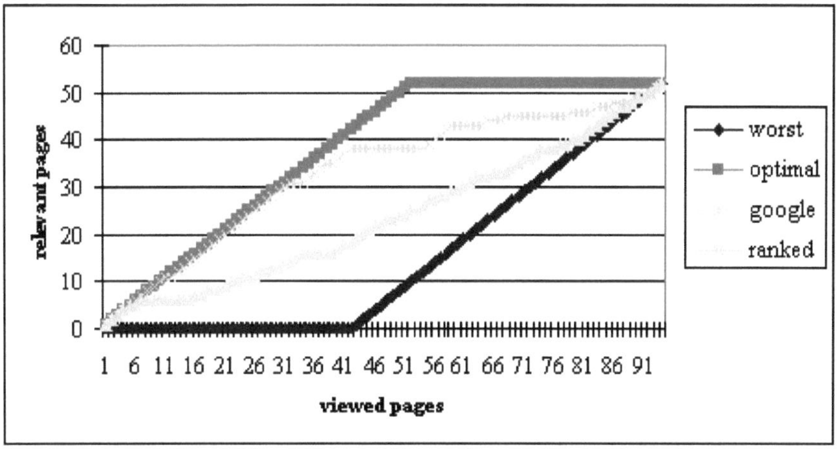


**Fig. 4.** Ranking for the concept *movie.*

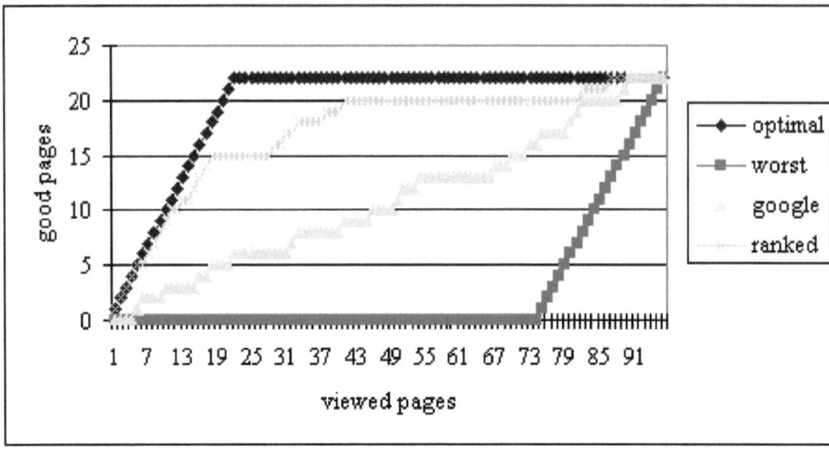


**Fig. 5.** Ranking for concept *travelling.*

## References

1. Agrawal R, Srikant R (1994) Fast Algorithms for Mining Association Rules. Research Report RJ9839, IBM Almaden Research Center, San Jose, CA.
2. Brin S, Motwani R, Page L, Winograd T (1998) The PageRank citation ranking: Bringing order to the Web. Stanford CS Technical Report.
3. Buchholz M, Pflüger D, Poon J (2004) Application of Machine Learning Techniques to the Re-ranking of Search Results. KI 2004: Advances in Artificial Intelligence, 27th Annual German Conference on AI, KI 2004, Ulm, Germany, September 20-24, 2004, Proc.
4. Frank E, Gutwin C, Nevill-Manning CG, PaynterGW, Witten IH (1999) KEA: Practical Automatic Keyphrase Extraction. Proc DL '99, pp. 254-256.
5. Pirolli P, Pitkow J, Rao R: Silk from a Sow's Ear (1996) Extracting Usable Structures from the Web. Proc of ACM SIGCHI '96, Vancouver, Canada, 118-125.

# Liver Perfusion using Level Set Methods

Sebastian Nowozin, Lixu Gu

Digital Medical Image Processing and Image Guided Surgery Laboratory, Shanghai Jiaotong University

**Abstract.** The family of Level Set Methods has been successfully used by scientists and practitioners for medical image processing. Image segmentation using active implicit contours is applied in 2D and 3D medical imaging, with the most popular methods being the Fast Marching Method and the Narrow band Level Set Method.

In this paper we apply level set segmentation to aid in automating the clinical challenge of measuring the contrast agent concentration in liver perfusion time series. For this, we apply implicit contour methods to time series of two-dimensional MRI images to yield accurate measurements of local image properties located relative to the shape of the liver across all images in the series.

Our results show that Level Set Methods can be used to provide the necessary segmentation shape data to reliably measure local image intensities positioned relative to this shape throughout a time series, where the location and shape of the object to be tracked changes.

## 1 Introduction

For certain illnesses related to the liver the blood flow to the liver has to be studied. By injecting a contrast agent into the patients body while taking MRI images in fixed time intervals, the concentration of the contrast agent can be studied while it flows through the patients body. A short time after the injection the contrast agent reaches the liver and the MRI images at that time can reveal important information about the blood supply condition of the liver for that patient. This information can lead to a more accurate diagnosis. The overall procedure is called *liver perfusion.*

Part of automating the above process can be modeled as a two dimensional registration problem, where the liver is the object of interest that is registered between the images. The more important problem is what to use as input to the registration mechanism. We use the segmented liver shape and now describe previous work on segmentation algorithms for two dimensional MRI images.



*G. Hommel and Sheng Huanye (eds.), Human Interaction with Machines,* 177-188.

An excellent discussion of medical image segmentation algorithms including level set methods is Suri et al. (Suri *et al.*, 2001a) and (Suri *et al.*, 2001b). The introduction of level set techniques into the field of medical image segmentation is due to Malladi in 1995 (Malladi & Sethian, 1995). Malladi used the curvature and the gradient of the image convolved with a Gaussian as a potential field to guide the evolution of the level set function. Our segmentation approach is based on his work, using a refined speed function. We have not found literature concerning the automation of perfusion measurements by incorporating level set methods for segmentation.

A large part of the effort to improve level set segmentation has been focused on merging a-priory knowledge or regional statistics into the speed function. This is necessary as a simple non-regional speed function will let the level set isocontour leak over weak or partially non-existing boundaries. Ho et al. (Ho *et al.*, 2001) replace the propagation term with a force based on regional statistics and let adjacent regions compete for a common boundary. They demonstrate improved results for brain tumor segmentation in MRI. Suri used fuzzy classifications of regions in (Suri, 2000) to build a speed function incorporating shape, region, edge and curvature information and applied the resulting model successfully to brain segmentation.

The current method to perform liver perfusion measurements is manual, because the patient breathes throughout the series and the liver moves vertically in a coronal view. As such the position of the blood vessel to be studied has to be marked in every single image. Then the intensity of the MRI image at these positions is plotted over time and from this the concentration curve of the contrast agent is deduced. The curve is used for further diagnosis.

In this paper we describe an automated method to locate the perfusion area within the time series of two-dimensional MRI images and present first results of the method applied to perfusion series. We have not found prior methods in the literature specific to liver perfusion measurements. Our method is novel in that it (a) combines segmentation results with a simple and efficient registration scheme specific to liver perfusion, (b) requires no manual interaction after a manual initialization and (c) can deal with small errors in the segmentated shape.

# 2 Methods

Now we describe briefly the fundamental methods we use througout the segmentation process.

## 2.1 Fast Marching Method

The Fast Marching Method (FMM) is an algorithm to efficiently solve curve and surface evolution problems. Consider a closed curve that evolves under a fixed-sign normal speed $F(x, y)$ dependent only on the position $(x, y)$ in the computational domain. The curve either expands outward all the time or moves inward all the time and once a point has been crossed by the curve it will never be crossed again. Then, the Eikonal equation can be given as

$$|\nabla T| F = 1$$

where $T(x, y)$ is the arrival time at which the curve or surface crosses the given point $(x, y)$. The Eikonal equation states the relationship that the gradient of arrival time is inversely proportional to the speed of the surface (Malladi & Sethian, 1996). The Fast Marching Method explicitly constructs the solution $T(x, y)$ for all points $(x, y)$ in the domain. The complexity for $N$ points is $O(N \log N)$ and the algorithm generalizes naturally to three or more dimensions. The original paper about the Fast Marching Method is from Sethian and Adalsteinsson (Adalsteinsson & Sethian, 1995). Since then, the Fast Marching Method has been intensively studied, the most detailed study is given by Sethian himself in (Sethian, 1998).

## 2.2 Narrow band Level Set Method

The level set method deals with the representation and evolution of closed interfaces, such as curves and surfaces. In the level set framework, the interface is *embedded* into a function of a dimensionality one higher than the original interface. Most often the signed distance function is used as embedding function, where the zero crossings of the function values represent the original interface. This implicit representation has many advantages. For one, geometric properties such as the local curvature $\kappa$ and the normal vector $\vec{N}$ can be easily determined. Most importantly, the evolution of the curve under the initial value problem

$$\phi_t + F|\nabla\phi| = 0$$

where $\phi$ is the embedding function, $\phi_t$ its derivate over time and $F$ is the speed function, can be solved iteratively. The speed function $F$ can be dependent on time dependent properties of the interface, such as the local curvature, the enclosed area and other global shape properties.

The level set method was first introduced in (Osher & Sethian, 1988), but since then the flexibility of the level set framework and the availability of high order accurate discretization methods have lead to adoption of the level set method in a large number of scientific fields, such as into computational physics, computer vision, image processing and medical image segmentation (Sethian, 1998), (Osher & Fedkiw, 2003).

The full level set method (Osher & Sethian, 1988) represents the curve on a discretized computational domain by assigning each point $(i, j)$ in the domain a corresponding function value $\phi_{i,j}$ . If one is interested only in the evolution and representation of the curve itself as opposed to the entire computational domain, it is enough to define $\phi$ within only a small boundary around the ZLC. This reduces the computational complexity significantly and is known as the Narrow band Level Set Method.

## 3 Segmentation

We developed a segmentation process for liver shape segmentation employing three steps. The first step locates a seed point for the segmentation in every image. The second step applies the Fast Marching Method to the original image at the given seed point to yield a first approximation of the liver shape. In the third step, this shape is used to initialize a level set segmentation step, which introduces a curvature reducing term to improve the segmentation results and repair local irregularities in the segmented shape. We now describe the steps in detail.

### 3.1 Locating the seed point

The seed point marks the initial curve position in the image. The initial curve is then evolved to segment the shape of the liver. The location of the segmentation seed point is semi-automatic. For one image, the radiologist manually marks the segmentation seed point. For all other images it is located using the following robust, but specific method.

In the MRI perfusion series, there are two patterns, (a) the top of the liver shape is well contrasted to the background, and (b) as the patient breathes throughout the series, the liver moves only vertically. Combining these patterns, we construct a simple method to locate the seed point: for every image a gradient magnitude is extracted and slightly mean smoothed inside a vertical strip at the horizontal position of the initial seed point. The maximum gradient magnitude is located in the strip, and a fixed $\Delta y$ is added to its vertical position. The value of $\Delta y$ is determined once in the original image, for which we already know the seed point location.

## 3.2 FMM Segmentation Step

The FMM segmentation step takes as input all the images and a single seed point for each image. As result, a segmented liver shape is returned as bitmap of the same dimension as the input image. For each pixel in the map which has a positive truth value the pixel is thought to belong to the liver shape. The FMM algorithm is fairly straightforward and the only flexible part is the definition of the speed function to use and the stopping criteria. The speed function determines the propagation speed in normal direction for any point in the computational domain. The stopping criteria tells us when to stop the segmentation.

The speed function $F_{FMM}$ we use for the FMM segmentation step is a thresholded variation of the well known speed function used in (Malladi & Sethian, 1995). We first define

$$F_{base}(x,y) = \frac{1.0}{1.0 + \left|S_k \cdot \nabla\left(G_\sigma * I_{x,y}\right)\right|^{S_p}}$$

which is the interface propagation speed in normal direction based on the gradient magnitude image. The gradient image $\nabla\left(G_\sigma * I_{x,y}\right)$ is the Sobel approximated gradient of the original image $I$ convolved with a Gaussian of width $\sigma$. $\sigma$, $S_k$ and $S_p$ are constants and we had good results using $\sigma = 3.0$, $S_k = 13.0$ and $S_p = 2.0$ for the series examined. The speed image is generated from $F_{base}$ by using it in the thresholded speed function $F_{FMM}$, which is defined

$$F_{FMM}(x,y)=\begin{cases}F_{base}(x,y) & F_{base}(x,y)\geq S_t\\ 0 & F_{base}(x,y)<S_t\end{cases}$$

with $S_t$ being the threshold value. We used a quite high value of $S_t = 0.6$, because the FMM cannot incorporate curvature dependent information. A segmentation using an unthresholded speed function could leak out of the liver shape where the gradient is locally weak. By using a cautious threshold we limit the risk of leaking in exchange for a higher probability to not cover the entire liver shape in the first segmentation step. By using the more powerful and robust levelset evolution method in the next step, we overcome this defect. The stopping criteria for the FMM segmentation is simply a constant of the area size, which corresponds to the number of iterations of the FMM. The value is configurable but we choose a default of 1500 elements, which yielded good results.

### 3.3 Levelset evolution step

The input to the levelset evolution step is the first rough FMM segmentation result. In this step we incorporate a regularizing curvature term to smooth the shape, remove holes and refine the segmentation result. The result is the level set signed distance map for the entire computational domain.

We use a narrow band extending 6 elements into both directions. The speed function used is the original function used above plus a curvature regularizing term. It is

$$F(x,y,t)=\alpha\cdot-\kappa_{x,y,t}+F_{base}(x,y)$$

with $\alpha = 0.4$ constant, $\kappa_{x,y,t}$ being the local curvature at the point $(x,y)$ at time $t$. The negative curvature term removes any small local irregularities the FMM segmentation has left over, such as sharp corners and single non-shape points within the shape due to noise in the original image. Globally it leads a smoother overall shape. We use a fixed number of evolution time steps of $\Delta t = 0.2$. To improve the results of the FMM segmentation, we found 80 time steps a good value. Finally, to simplify the perfusion area localization, the narrow band levelset function is extended to a full levelset in the entire domain by redistancing from the shape contour in the narrow band.

## 4 Perfusion Area Localization

We now introduce a new method to locate the perfusion area in each image given the segmented liver shape. In this method, the radiologist first marks the perfusion area in one image. Afterwards, this area is automatically located in all remaining images.

Our method is based on the following five observations. First, the movement of the liver is constrained to mainly vertical movements with only small horizontal movement. Second, shape of the segmented liver changes only slightly throughout the series. Third, rotation relative to the body is minimal and can be ignored. Fourth, the segmentation quality is not equal across the entire shape but the best at the top half of the liver, due to a strong gradient response there, while the weak gradient response in the lower half of the liver leads to more variation throughout the series. Fifth, when the contrast agent reaches the perfusion area a strong gradient response appears, which changes the segmentation result locally, up to the case of where the perfusion area is not considered part of the liver shape anymore. We now describe the details of the method. The idea is to anchor a coordinate system at each segmented shape which can be used to estimate a point location within and nearby the liver shape across all the images.

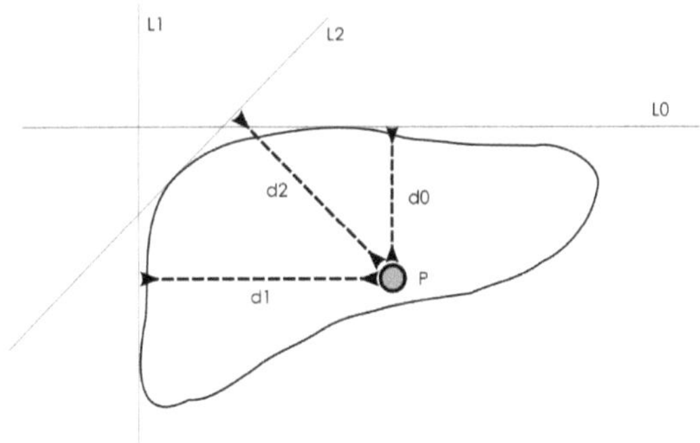


**Fig. 1.** A point $P$ defined by a three element distance vector $(d_0, d_1, d_2)$ relative to the lines $L_0$, $L_1$ and $L_2$.

Consider the point $P$ in figure 1. Assume for now, we know for sure the absolute position of $P$ in the image and that we have a good segmentation result of the liver shape. Then, we define a set of non-parallel lines $L = \{L_0, L_1, \ldots, L_n\}$. For each line $L_k$ we determine the following for the image $I$ : 1. the distance $d(P_{x,y}, L_k, I)$ of every point $P_{x,y}$ within the segmented liver shape to $L_k$ and 2. the shortest distance $d_s(L_k, I)$ among all $d(P_{x,y}, L_k, I)$. Then, any point within the liver in an image $I$ can be represented as a *line relative distance vector* $D_{P_{x,y}}(I)$:

$$D_{P_{x,y}}(I) := \left(d(P_{x,y}, L_0, I) - d_s(L_0, I), \ldots d(P_{x,y}, L_n, I) - d_s(L_n, I)\right)$$

Using the above model, what remains to be discussed is how to find suitable lines to build the distance vector from and how to transfer from a given distance vector and the lines back to an absolute coordinate. The first question can be answered by considering the fourth observation made above. We use lines which always have their minimum distance point at the boundary of the shape where the segmentation result is of good quality. For example by using a horizontal line above the liver as distance measurement line, the resulting component in the distance vector will accurately reflect the relative position to the top of the liver across all the images because there is a strong gradient response at the top of the liver. Similarly, the top left part of the liver is always well segmented and by fitting a diagonal line with a 45 degree angle, we yield another good element for the distance vector.

To obtain an absolute coordinate given $D_P(I_p)$ for the initial radiologist-marked image $I_p$ and the segmentation results for all other slices, we determine the position within or nearby the segmented shape that minimizes an error term. Consider a two dimensional coordinate system, where two non-parallel lines are enough to define a base and any additional lines are redundant. This redundancy can be used to define an error term $\varepsilon$ which describes how much the distance vector for a point $(x_0, y_0)$ in image $I_0$ diverges geometrically from the distance vector of the original known perfusion area point $(x_p, y_p)$ in the radiologist-marked image $I_p$:

$$\varepsilon(x_0, y_0, I_0) := \left|D_{P_{x0,y0}}(I_0) - D_{P_{xp,yp}}(I_p)\right|$$

The *error minimizing point* $(x, y)$ in the image $I$ is one of the points that minimizes $\varepsilon(x, y, I)$. To find this point we test all points $(x, y)$ for which the level set value is below some small positive threshold value $t$ and keep the minimal error point. Because the embedding function used with the levelset method is the signed distance function, a small positive value

$t$ allows for points nearby the segmented shape to be found, which is necessary because of the fifth observation made above.

After all the minimizing error points are located in the images, the perfusion area is simply a circle area centered around the point in each image. Because the liver shape is deformed only slightly throughout the series and the circle area is invariant to rotation, it is a simple but sufficient approximation. For each image, the resulting perfusion intensity is the mean value of all the MRI intensity values within the circle area[1].

## 5 Experimental Results

In this section we evaluate our implementation of the proposed method. We created a new implementation of the Narrow band levelset method and the FMM, written in C. A prototype GUI was written in C# using the Gtk# toolkit. We evaluate the performance of the proposed method on three perfusion series. The series consists of two-dimensional 256x256 MRI images taken with a GE Medical Systems Genesis Signa system at the Shanghai First People Hospital. They show the patients abdomen in coronal view. The imaging parameters are the following: slice thickness 15.0, repetition time 4.7, echo time 1.2, magnetic field strength 15000, flip angle 60 degrees. The images were converted to the PNG file format using medcon[2] with the following options: `medcon -e 1 0 -fb-dicom -c png -f *.dcm`. The first series consists of 240 images, the second series of 59 images and the third series of 200 images.

The evaluation has been performed using our prototype GUI. In the GUI, one particular good image of the series is selected and the seed point and the perfusion area is manually marked. Afterwards the segmentation and perfusion measurement process is started, which produces the intensity curve with its minimal error term values.

---

[1] The unit of this intensity is arbitrary and to obtain the contrast agent concentration from it requires more effort and also depends on which agent is used.

[2] Medcon, medical image conversion tool, http://xmedcon.sourceforge.net/

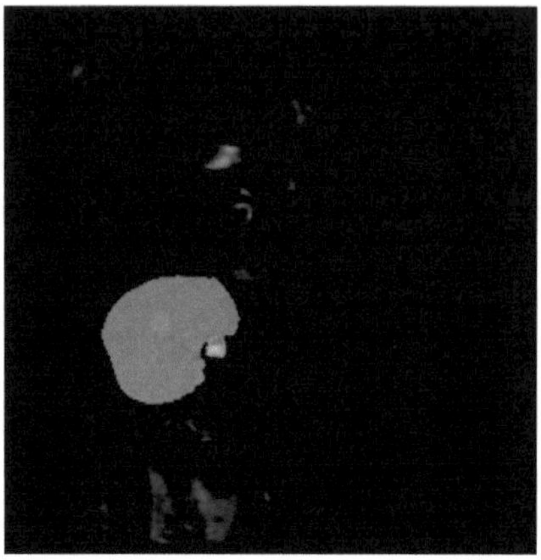

**Fig. 2.** A processed example image from the first series. For this case, the marked perfusion area is outside the segmented liver shape, as explained in the text.

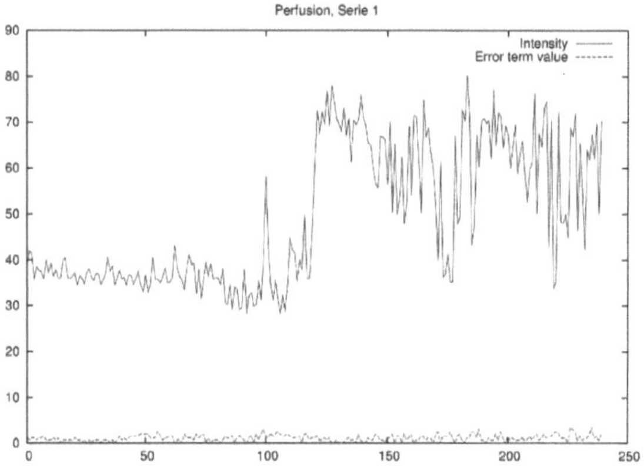


**Fig. 3.** The perfusion area intensity time curve and the corresponding error term values of the first perfusion series (240 images).

In figure 2 a typical processed image is shown. The seed point and the perfusion area is marked, and the segmented liver shape is drawn as overlay over the original image. The output intensity curve for the first series is shown in figure 3. Around the $120^{th}$ image, when the contrast agent reaches the liver, a strong response can be seen clearly. The error term values are also shown. They do not reflect the absolute accuracy of the located liver perfusion area. Instead they allow to give a relative comparison between the individual images.

**Table 1.** Error term comparison for the three example perfusion series.

| Series | Images | Error mean | Error maximum | Error standard deviation |
|---|---|---|---|---|
| Series 1 | 240 | 1.1401 | 3.5355 | 0.7348 |
| Series 2 | 59 | 2.9098 | 5.5227 | 1.5197 |
| Series 3 | 200 | 2.5054 | 7.0356 | 1.6348 |

In table 1 we analyze the error terms within the three series. Interpreting the error values as the geometric distance from an optimal fit, the low mean error values for all the series indicates a good perfusion area localization success, which is confirmed by manually inspecting the processed images. For the runtime performance evaluation, the first series has been used. The system is a Pentium-M 1500Mhz, 512Mb RAM system.

**Table 2.** Runtime performance measurements.

| Step | time | time per image |
|---|---|---|
| Locating seed point | 44s | 0.183s |
| FMM and levelset liver segmentation | 443s | 1.845s |
| Locating perfusion area | 23s | 0.095s |
| Total | 510s | 2.125s |

## 6 Discussion and Conclusions

We proposed and evaluated a combined segmentation and registration method to extract local image intensity from the liver. For this we developed a simple and robust registration method which maps the perfusion area to a minimum error fit coordinate in each image.

The results still have to be verified for its clinical value by radiologists. We confirmed that the combined FMM and Narrow band Level Set Methods based segmentation approach is computationally efficient.

In the near future we will seek for confirmation and advice from radiologists to improve on this results. Also, further research is needed to draw conclusions about the performance of the proposed method in case the segmented objects have a more complex shape or vary considerably across the images in the series, such as the heart. In our method the quality of the measurement increases with improved segmentation results and we consider incorporating a model-driven segmentation approach to improve accuracy.

## References

1. Adalsteinsson, D., & Sethian, J. 1995. A Fast Level Set Method for Propagating Interfaces.
2. Ho, Sean, Bullitt, Elizabeth, & Gerig, Guido. 2001 (11). Level Set Evolution with Region Competition: Automatic 3-D Segmentation of Brain Tumors. Tech. rept. TR01-036.
3. Malladi, R., & Sethian, J. A. 1995. Image Processing via Level Set Curvature Flow. *Proc. Natl. Acad. of Sci.*, **92**(15), 7046-7050.
4. Malladi, R., & Sethian, J. A. 1996. An O(N log N) algorithm for shape modeling. *Proceedings of the National Academy of Sciences*, **93**(Sept. 25), 9389-9392.
5. Osher, Stanley, & Fedkiw, Ronald. 2003. *Level Set Methods and Dynamic Implicit Surfaces*. Springer.
6. Osher, Stanley, & Sethian, James A. 1988. Fronts Propagating with Curvature-Dependent Speed: Algorithms Based on Hamilton-Jacobi Formulations. *Journal of Computational Physics*, **79**, 12-49.
7. Sethian, J. 1998. Level Set Methods and Fast Marching Methods: Evolving Interfaces in Computational Geometry. Cambridge University Press.
8. Suri, J., Liu, K., Singh, S., Laxminarayana, S., & Reden, L. 2001a. Shape Recovery Algorithms Using Level Sets in 2-D/3-D Medical Imagery: A State-of-the-Art Review.
9. Suri, J.S. 2000. Leaking prevention in fast level sets using fuzzy models: an application in MR brain. *IEEE EMBS International Conference on Information Technology Applications in Biomedicine, 2000. Proceedings*, 220-225.
10. Suri, J.S., Setarehdan, S.K., & Singh, S. 2001b. Advanced Algorithmic Approaches to Medical Image Segmentation: State-of-the-Art Applications in Cardiology, Neurology, Mammography and Pathology. Springer.

MIX
Papier aus verantwortungsvollen Quellen
Paper from responsible sources
FSC® C105338

If you have any concerns about our products,
you can contact us on
**ProductSafety@springernature.com**

In case Publisher is established outside the EU,
the EU authorized representative is:
**Springer Nature Customer Service Center GmbH**
**Europaplatz 3, 69115 Heidelberg, Germany**

Printed by Libri Plureos GmbH
in Hamburg, Germany